# History of Physics

The Springer book series *History of Physics* publishes scholarly yet widely accessible books on all aspects of the history of physics. These cover the history and evolution of ideas and techniques, pioneers and their contributions, institutional history, as well as the interactions between physics research and society. Also included in the scope of the series are key historical works that are published or translated for the first time, or republished with annotation and analysis.

As a whole, the series helps to demonstrate the key role of physics in shaping the modern world, as well as revealing the often meandering path that led to our current understanding of physics and the cosmos. It upholds the notion expressed by Gerald Holton that "science should treasure its history, that historical scholarship should treasure science, and that the full understanding of each is deficient without the other." The series welcomes equally works by historians of science and contributions from practicing physicists.

These books are aimed primarily at researchers and students in the sciences, history of science, and science studies; but they also provide stimulating reading for philosophers, sociologists and a broader public eager to discover how physics research – and the laws of physics themselves – came to be what they are today.

All publications in the series are peer reviewed. Titles are published as both print- and eBooks. Proposals for publication should be submitted to Dr. Angela Lahee (angela.lahee@springer.com) or one of the series editors.

Aldo Treves · Pasquale Tucci
Editors

# Enrico Fermi, Atomic Physics Lectures

Società Italiana di Fisica

Springer

*Editors*
Aldo Treves
Department of Science and Technology
University of Insubria
Como, Italy

Pasquale Tucci
Department of Physics
University of Milan
Milan, Italy

ISSN 2730-7549 ISSN 2730-7557 (electronic)
History of Physics
ISBN 978-3-031-68689-4 ISBN 978-3-031-68690-0 (eBook)
https://doi.org/10.1007/978-3-031-68690-0



This Springer imprint is published by the registered company Springer Nature Switzerland AG
The registered company address is: Gewerbestrasse 11, 6330 Cham, Switzerland

# Foreword

The Italian Physical Society (SIF) considers very important to have easily available all the documents concerning the research of Fermi on quantum mechanics and nuclear physics. The scientific work of Fermi is so outstanding that no parts of it should be neglected. In fact, as it was well recognized internationally from the start of his research activity, Fermi has made important discoveries and has provided in depth insights in all the physics questions he has addressed and investigated. His scientific production is indeed very rich. In addition, as illustrated in a recent book published by SIF "Il Grande Fermi" (edited by edited by Luisa Cifarelli and Giuseppe Pelosi) Enrico Fermi is a great public figure and his name is found in many things (prizes, schools, streets, stamps, movies, etc...).

It is then not surprising that the SIF enthusiastically decided to support the proposal of Aldo Treves and Pasquale Tucci to prepare an English text containing the lectures given by Fermi in 1949 in Italy. These lectures, that were originally published only in Italian, were given by Enrico Fermi as "Atomic Physics Conferences" in Rome and Milan.

To have a book published jointly by SIF and Springer on these lectures is a special initiative taking place at the occasion of the 70th anniversary of his death. The published material has used and commented the notes from Professors and Physics Assistants from the Universities of Rome and Milan who participated in these lectures. The topics presented in this book, well organized by the authors, are interesting not only for historical reasons but also because they are still very timely. In fact, many aspects of them are addressed nowadays with modern experimental and theoretical techniques.

I thus expect this book to attract the curiosity of readers interested in learning on the roots of the modern basic research and of the technical developments impacting our lives. The pedagogical skill of Fermi is well underlined by the authors. We are aware that without memories it is difficult to design the future!

On behalf of SIF I like to thank, in addition to Aldo Treves, Pasquale Tucci, also the other contributors to this volume. Their work will be surely much appreciated by the readers.

Angela Bracco
President of the Società Italiana di Fisica (SIF)
Department of Physics
University of Milano
Milan, Italy
angela.bracco@sif.it

# Preface

In 1949 on the occasion of the 150th anniversary of the invention of the Volta battery an International Conference on Cosmic Rays was held in the Villa Monastero (Varenna). Fermi, returning to Italy for the first time for 11 years, attended the conference. He surely talked to his old Italian friends about the period of secrecy and poor communication forced by his participation in the Manhattan Project for the construction of the atomic bomb. But he had not lost his special interest in basic research, his proverbial ability to connect experimental results with physical theories and his wishing to communicate to a vast audience the new achievements in physics. It is not surprising, therefore, that he accepted the invitation of the Accademia dei Lincei and the Donegani Foundation to give a series of lectures on the development of atomic physics: six lectures at the Physical Institute of the University of Rome, and three in Milan in the assembly hall of the Montecatini Society.

The lectures were mainly aimed at physicists who wanted to be informed and trained about update research in atomic physics, from which they had been excluded during the war. It was no coincidence that the lectures were edited by professors and assistants from the Universities of Rome and Milan who would be among the protagonists of the rebirth of Italian and European physics after the Second World War: S. Sciuti, L. Mezzetti, E. Pancini, N. B. Cacciapuoti, G. Morpurgo, C. Salvetti and P. Caldirola. We have added at the end each Fermi's lecture a short biographical note on the editors of the Italian edition, first published by the Accademia dei Lincei.

With impressive speed the lectures were published by the Accademia dei Lincei in 1950. They are inserted in Collected Papers (1962 and 1965),[1] and two of them (the first "The Elementary Particles" and the seventh "The Neutron") appear in the book edited by Vincenzo Barone.[2] The Lincei volume is impossible to find and the Collected Papers are aimed at a specialist audience. The nine lectures, originally

[1] Fermi, Enrico (1962 and 1965). Collected Papers (Chicago-Rome: The University of Chicago Press—Accademia Nazionale dei Lincei, 1962 and 1965), 2 vols.

[2] Barone, Vincenzo (ed.) (2009). Enrico Fermi. Atoms Nuclei Particles (Turin: Bollati Boringhieri, 2009).

published in Italian, are presented for the first time in an English translation. The mathematical notation has not been modified.

The nine lectures are an important documentation of issues that Fermi judged appropriate to submit to the reborn community of Italian and European physicists after the physical and moral destruction of the war. In addition to the numerous mentions of the results of his group in Rome (see Lectures 7 "The Neutron)", Fermi extensively quoted the results of Italian physics after his departure. In particular, in the first lecture ("Elementary Particles") his focus was on the tracks of cosmic radiation on Wilson chambers and photographic plates whose "beauty" he seems fascinated by, for which the contribution of G. Occhialini was decisive. He emphasized the importance of the experiment of Conversi, Pancini, Piccioni "carried out in a cellar during the period of the resistance", which demonstrated that the meson, later called $\mu$ meson, could not be the Yukawa particle. It would be identified with the $\pi$ pion that decays into a $\mu$ meson and Fermi reports the famous photographic emulsion images produced by Powell and Occhialini. Fermi began the lecture with a list of nine particles which he considered to be "elementary". Among them the positive and negative pions appear, but not the neutral one, which had not yet been discovered. The neutrino is in the group and it is labelled as "possibly massless". The lecture contained an exemplary clear exposition of the Yukawa theory, with an elementary deduction of the mass of the meson (pion).

In the second lecture on Elementary Particles, in the interpretative part, which was intended to report some initial ideas, together with Wheeler's proposals, he quoted Puppi's contributions. In the last lecture "Dirac's monopole" Fermi asked who could discover the monopole and he claimed "that this could happen in a country, like Italy, where studies on cosmic rays are so flourishing."

The structure of each lecture fully reflects Fermi's style: formulation of the problem, experimental observational evidence, theoretical interpretation. The latter in the lectures is reported in descriptive form, with very limited mathematical support. In the third lecture "Theories on the Origins of the Elements" the starting point was the "cosmic distribution of the elements": in it he described the evidence from meteorites, stellar spectra etc. This is an example of Fermi's interest in topics that we would now call astrophysical, and which would also be developed thanks to the collaboration with Chandrasekhar that would lead to the fundamental work on models of acceleration of cosmic radiation. Fermi referred to Gamow's theory. He spent a few words on him, which will certainly have amused the public and will make historians smile: "Gamow, who, being as everyone knows a joker, has associated himself with two other physicists Alper and Bethe" so as to remember the first three letters of the Greek alphabet. The theory was based on the observation "of the possibility of forming elements, even when the temperature and pressure do not assume such astonishing values as those mentioned above, provided that the elements are supposed to be formed through successive additions of neutrons." The attention that Fermi paid to Gamow's work is certainly remarkable. It was from it that the idea of the cosmic microwave background was born, which would revolutionize cosmology. In the fifth lecture "Nuclear Orbits" the problem was posed as follows: to evaluate the binding energy of a nucleus consisting of Z protons and N neutrons. He then moved on to the experimental results, mentioning The Magic Numbers, the Seaborg results with their extension of the alpha radioactivity table. Fermi proposed two interpretative models: the first was the nuclear liquid drop model, of which he made a very clear synthetic presentation. The second was the attempt to replicate the atomic model, in the case of the nucleus, which he obviously preferred. Fermi mentioned the difficulties and arguments for which the model should not work, and then to his satisfaction, and almost as a "happy ending" for the audience, he reported the successes of the model.

In the sixth lecture "New Developments in Electrodynamics", which was for a rather specialistic audience, Fermi described Lamb's spectroscopic result on hydrogen levels and other experimental results that require a specific and detailed description of field-electron interactions. This was addressed with a perturbative theory which yielded singularities but these needed to be cured. This is a highly complex subject, of which Fermi gave a qualitative description, introducing the first advances in the theory of renormalization. Fermi was not satisfied with the theory, and one might wonder if he would be satisfied with its current state, after seventy years. The words with which he closed the conference seem to have a prescient meaning. "There is a great movement at the moment: almost all young theoretical physicists are anxious about field theories and nothing but field theories, perhaps a little too one-sided: but only the future will tell."

The book contains two essays by very well known physicists and historians of physics who have contributed greatly to the historiography on Fermi, in line with the well known statement by Amaldi that the history of physics should be done by physicists.

Vincenzo Barone has focused on an important aspect of Fermi's activity, aiming to inform non-specialists on the progress of modern physics. Barone, who has dedicated a book on this subject (See footnote 2), points out that simplification is the fil rouge followed by Fermi in both his scientific and outreach papers. "Simplicity was for Fermi both a method of research and a style of communication". In the period before 1938, besides his research in physics, Fermi's activity was addressed to various audiences, scientists not specialists in physics, high school students, the general public, and political elites, which at the time were in the fascist party, see e.g. Fermi's article, mentioned by Barone, on the nascent nuclear physics published in the superfascist journal *Gerarchia.*

The paper by Guerra and Robotti is an excursus on Fermi's life and scientific achievements, using archival material, from the Domus Galileiana, the Amaldi Archives in Rome University and the Fermi Archives in the University of Chicago. Fermi's exceptional intellectual capacity, which appeared from his infancy, is contrasted to his behavior as a good decent citizen, family man and father. Guerra and Robotti concentrate on a number of subjects where Fermi's work was fundamental. An example is the competition with Heisenberg for the construction of the first nuclear reactor and the reasons why Heisenberg did not achieve this goal.

Aldo Treves
Department of Science and Technology
University of Insubria
Como, Italy
aldo.treves@uninsubria.it

Pasquale Tucci
Department of Physics
University of Milan
Milan, Italy
pasquale.tucci@unimi.it

# Acknowledgements

This volume is published under the auspices of the Accademia dei Lincei, which organized the Lectures with Fondazione Donegani and put in print, in 1950, the Italian edition of the Lectures.[3]

In editing this book we wish to express our gratitude to Ms. Barbara Ancarani of the Italian Physical Society (SIF), for her continuous help and suggestions, and for coordinating the adjustment of the figures, images and graphs.

We are grateful also to Ms. Marina Forlizzi and to Ms. Barbara Amorese of the Springer Italy for their help on the English editing of Fermi's lectures.

English translation of Fermi's lectures was revised by Ms. Christine V. Pennison with not only linguistic competence.

[3] Fermi, Enrico (1950). *Conferenze di Fisica Atomica. Raccolte da professori ed assistenti di Fisica delle Università di Roma e Milano* (Roma: Accademia Nazionale dei Lincei, 1950).

# Contents

# Enrico Fermi. Life

**Francesco Guerra and Nadia Robotti**

## 1 Enrico Fermi. Life

### *1.1 Beginning*

Enrico Fermi was born in Rome on September 29, 1901, to Alberto, an employee of the Ministry of Communications, and Ida De Gattis teacher, both of north Italy descent, as the last of three siblings, Maria (1889–1959) and Giulio (1900–1915). After attending the ginnasio-liceo classico Umberto I in Rome (now Pilo Albertelli), skipping a year, in 1917 he won a position as a student at the Scuola Normale Superiore in Pisa and enrolled in the graduate course in physics at that university. From an early age his extraordinary predispositions toward scientific subjects become evident.

Significant in this regard is a notebook from second grade primary school, which he took with him when he moved permanently in 1939 to America and now preserved at the Chicago Library of the University of Chicago, along with documents relating to his "American" period. It shows that Fermi at the age of six was already very mature in his compositions, mastery of language and calligraphy, and above all very knowledgeable about technical-scientific and even practical matters. (See Figs. 1, 2 and 3).

These predispositions are reinforced by important readings chosen independently by Fermi himself, but also recommended by family friends and by one of his High School Professors, Filippo Eredia (1877–1948), a physicist from Catania who moved to Rome and later became a famous Italian meteorologist, geophysicist and academician. This self-taught preparation will ensure that when he faces the National

F. Guerra (✉)
Centro Studi e Ricerche Enrico Fermi; Università degli Studi di Roma "La Sapienza", Rome, Italy
e-mail: francesco.guerra@roma1.infn.it

N. Robotti
Centro Studi e Ricerche Enrico Fermi, Università degli Studi di Genova, Genoa, Italy

 
A. Treves and P. Tucci (eds.), *Enrico Fermi, Atomic Physics Lectures*, History of Physics,
https://doi.org/10.1007/978-3-031-68690-0_1

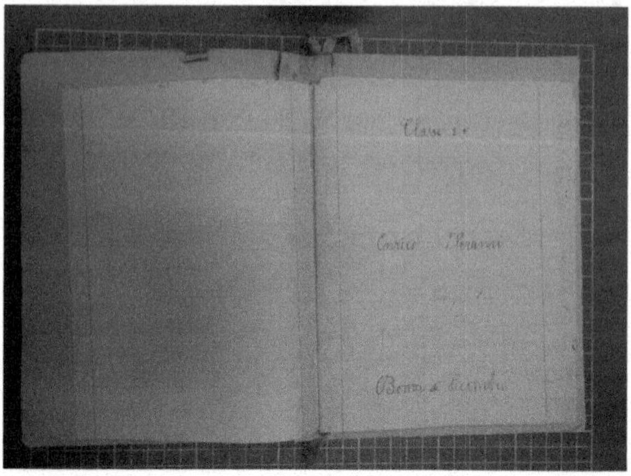

**Fig. 1** First page of the Fermi notebook used during his second grade primary school in Rome (Chicago Library. Chicago)

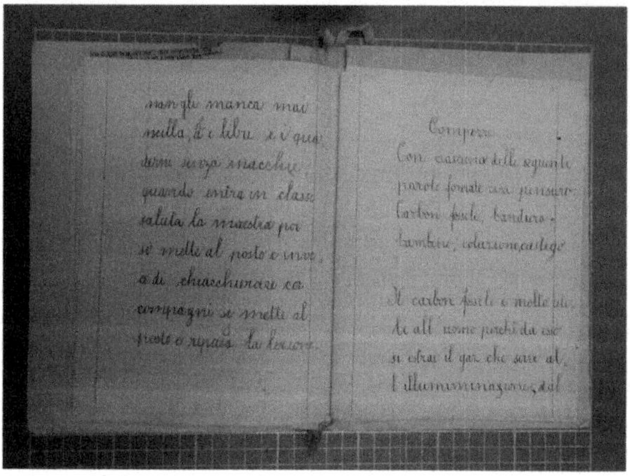

**Fig. 2** Fermi notebook used during his second grade primary school in Rome (Chicago Library. Chicago)

Competition for admission to the Scuola Normale Superiore in Pisa, he will place first, making himself noted for his uncommon preparation. It should be noted that one of the commissioners of the Competition for admission to the Scuola Normale will be precisely, along with the Mathematician Giulio Pittarelli (1852–1934) and the Biologist Federico Raffaele (1862–1937), Filippo Eredia.

While in Pisa, he broadened and enriched his scientific preparation by extending it toward the New Physics, as is evidenced by a notebook, dating from 1920, in which Fermi gives a summary, made by himself, of some of the main and most recent articles that came out in the international press concerning Relativity and Quantum Mechanics. We find, for example, articles in the names of Bohr, Einstein, and Sommerfeld, but among them we also find three still unpublished articles, under the name of "E. Fermi," written precisely by Fermi and which will later be published and concerning respectively a new arrangement for the study of high frequency spectra, the absorption of sound in gases, the dynamics of a rigid system of electrical charges.

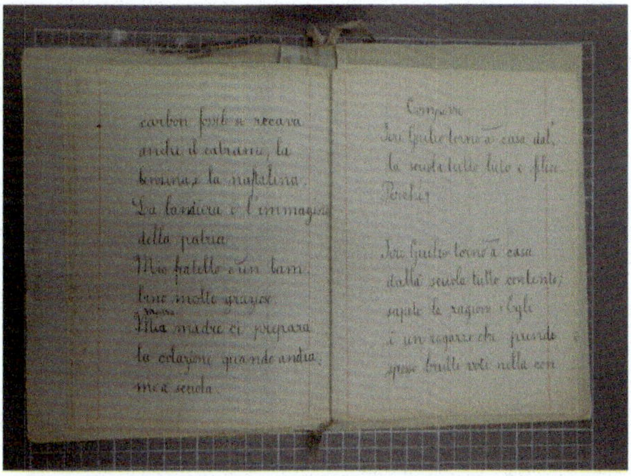

**Fig. 3** A page of the Fermi notebook used during his second grade primary school in Rome (Chicago Library, Chicago) (2)

**Fig. 4** Cover of the Fermi notebook (1921)

This notebook is preserved at the Domus Galilaeana in Pisa, along with almost all original documents related to Fermi's research activity in Italy. (See Figs. 4 and 5).

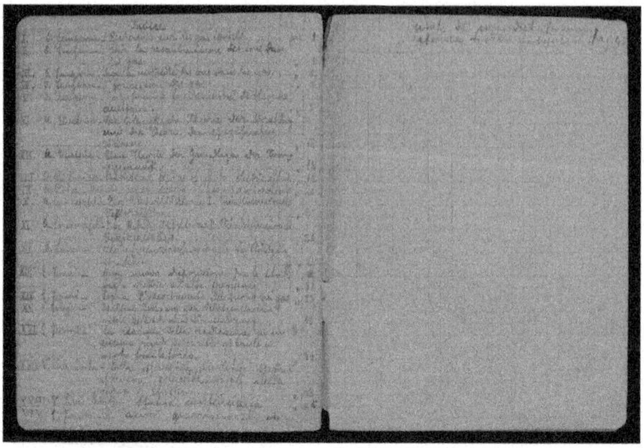

**Fig. 5** Fermi notebook (1921) (Domus Galilaeana, Pisa)

In short, Fermi, as early as 1920, placed himself, in a metaphorical sense, among the Greats and was also noticed by the Pisa academic world. As he wrote to Enrico Persico from Pisa on January 30, 1920:

> At the Physical Institute I am little by little becoming the most influential authority. Indeed one of these days I am to deliver, in presence of several magnates, a lecture on the quantum theory of which I am always a propagandist ...

In order to earn his degree Fermi must take 16 exams. Of these, as can be seen from a certificate from the University of Pisa, issued on July 7, 1922 and preserved at the Chicago Library of University of Chicago, 13 are passed with a grade of 30/30 *cum laude*, 2 with a grade of 30/30 and one, "Freehand Drawing" with a grade of 24/30. When we see the electrical circuits designed by Fermi for his research on neutron-induced radioactivity, we will understand how well deserved was this poor grade awarded to him.

On July 7, 1922, he graduated, with honors, discussing an experimental thesis on the diffraction X-rays. Title of the thesis is "Studies over X-rays". Of this thesis Fermi was not at all pleased. As he wrote to his friend Enrico Persico on March 18, 1922:

> These days I have a lot to do for my thesis, which, by the way, has come a rubbish of the most solemn.

Despite this negative judgment, Fermi's thesis had its scientific merit. Although Fermi's idea to form a real monochromatic image of an X ray source was not original (it had been anticipated by Louis Georges Gouy in 1915) and its realization not easy, it was taken up and successfully applied eight years later by M. Pierucci, in Pisa, to build an X-ray collector, thus leading to the recognition in Fermi of a forerunner in this field of study.

A few days later, on July 7, he also received the Diploma of Habilitation from the Normal School, with a theoretical thesis entitled "A theorem of calculus of probability and some of its applications.", later published.

## 1.2 After Graduation

Having completed his studies in Pisa, upon his return to Rome, he had a decisive meeting with Orso Mario Corbino, Senator of the Kingdom, Professor of Experimental Physics and Director of the Institute of Physics at the Royal University of Rome. This meeting will be decisive for his future and also for that of Italian physics. Corbino, interested in the scientific development of his Institute, realizes that he had found in Fermi the right element.

Following the practices of the time, Fermi in 1923 was accepted in Rome into the Masonic Lodge "Adriano Lemmi," then of obedience to the Communion of Piazza del Gesù, the same as Corbino.

Fermi's academic career developed steadily and rapidly. In fact, in 1923, with the support of Corbino, who was also a member of the examiners, he obtained a scholarship for study abroad from the Ministry of Education. He then goes for a few months to Göttingen, to the Institute of Max Born. There he extends his network of acquaintances with the leading physicists of the time, but his stay is somewhat ahead of its time. In fact, the important research on the new quantum mechanics will not start until 1925 with Werner Heisenberg, and would later be extended in collaboration with Born himself, and Pascual Jordan.

Upon his return to Rome, Fermi held by assignment the Course in Mathematics for Chemists for the academic year 1923–1924. In the fall of 1924, taking advantage of a scholarship from the Rockefeller Foundation, he went to Leiden for three months, where he established excellent relations with Paul Ehrenfest, receiving stimulation from him for his involvement in problems in statistical mechanics.

The visit to Leiden by Fermi is carefully prepared by Vito Volterra, who is responsible not only for obtaining the approval of the Rockefeller Foundation for financial support, but also writes personally to H. A. Lorentz to facilitate contacts with Paul Ehrenfest, and also with the experiments of Heike Kamerlingh Onnes and Willem Hendrik Keesom. In his long and detailed reply, dated Haarlem February 25, 1924, Lorentz assures that Enrico Fermi will find in Leiden a cordial welcome, and extensively outlines the possibilities of scientific contacts, in agreement with Fermi's research interests. The visit to Leiden is planned initially for the duration of six months, with an overall support of $700. But important new developments are looming on the horizon, and the visit to Leiden is reduced to three months beginning September 1, 1924.

At the Historical Museum of Physics and Enrico Fermi Study and Research Center in Rome, it is preserved a very significant letter from Enrico Fermi, dated Rome March 23, 1924, and addressed to Ettore Bortolotti, Professor of Geometry at the Royal University of Bologna. Here Fermi thanks him for the cordial welcome and hospitality he received in Bologna and mentions two his further visits to Pisa and Florence. Clearly at the beginning of 1924 Fermi was exploring in person the various possibilities open to him for his future accommodation, considering, in addition to the natural solution Rome, where he was already covering a teaching position, also Bologna, Pisa, and Florence. Fermi was certainly in relative confidence with Ettore

Bortolotti, as as he was also a classmate in Pisa of his son Enea. In the letter, Fermi reveals a fact very important, which we report directly in his own words:

> In Florence I made some arrangements for my possible accommodation at that university for the coming year.

This is the first hint that Fermi's academic trajectory is about to touch Florence. Seen from the outside, it certainly appears that Antonio Garbasso, director of the Florence Physics Institute at Arcetri, manages to achieve a strategically decisive result, vis-à-vis Orso Mario Corbino in Via Panisperna, securing the presence in Arcetri of the most brilliant young physicist in Italy already widely established with a scientific production of the highest level.

Certainly the Florence offer was more convenient for Fermi at the time, while waiting for Corbino could perfect the optimal strategy to secure Fermi at Via Panisperna. In fact, Fermi received in Florence for 1924–1925 the two teaching assignments in Rational Mechanics, for the third year of the undergraduate courses in Mathematics and Physics and Mathematics, and in Mathematical Physics for the fourth year. These courses were also taken by students in the propaedeutic biennium of Engineering. The assignments are renewed for the following academic year 1925–1926. Being able to teach two courses very close to his research interests will have been a great satisfaction for Fermi, who was freeing himself from the tedious course in Mathematics for Chemists. It is no coincidence that Fermi a few years later, in drafting his curriculum vitae for confirmation on the Chair of Theoretical Physics in Rome, states that the courses taken on assignment in Florence had been "theoretical physics and rational mechanics."

Even then, for Fermi, Mathematical Physics was to be understood as Theoretical Physics. Unfortunately,we do not have the lecture records of the Mathematical Physics course, so that we can not see how Fermi had organized his topics. Of course, the course in Rational Mechanics, given its propaedeutic nature did not allow much freedom in the choice of topics. The period spent on the Arcetri hill is extremely fruitful for Fermi.

In fact in 1925 he obtained a free professorship (libera docenza) in Mathematical Physics, securing for himself a much stronger academic presence. It should be noted that Theoretical Physics did not yet exist as an official discipline. Moreover, it was during the Florentine period that Fermi achieved his first scientific result of absolute international value: the statistical formulation of systems of identical particles, which follow the Pauli principle. Enrico Fermi publishes his results in a series of papers, beginning with a Lyncean note, entitled. "On the Quantization of the Perfect Monoatomic Gas," which appeared in early 1926, and was presented by Member Antonio Garbasso at the February 7 meeting of the Academy, followed by an article in the German Zeitschrift für Physik, entitled "Zur Quantelung des idealen einatomigen Gases" sent for publication on March 26.

A few months later, independently, Paul Dirac arrives at similar results, set forth in a paper published in the Proceedings of the Royal Society on Oct. 1 of 1926, but sent for publication on August 26, in which Fermi is not cited at all. But the Fermi's reaction is immediate. In a polite but firm letter under the letterhead of the Istituto di Fisica in Rome, dated October 25, 1926, and thus sent only a few days after the publication of Dirac's article, the truth of the facts is reestablished. (See Fig. 6) Fermi writes:

> Dear Sir! In your interesting paper "On the Theory of Quantum Mechanics" (...) you have put forward a theory of the Ideal Gas based on Pauli's principle. Now a theory of the ideal gas that is practically identical to yours was published by me at the beginning of 1926 (...). Since I suppose that you have not seen my paper, I beg to attract your attention on it. I am, Sir, Your Truly (signed) Enrico Fermi.

We do not have the text of Dirac's reply, but he certainly accepts the clarification of Fermi, so much so that in a later work he proposes to call Fermi's statistics the statistical distribution involved. History has crystallized the designation "Fermi-Dirac statistics."

Moreover, the corresponding particles have been called "Fermions." Without the letter of Fermi, perhaps the term "Dirachions" would have been arrived at. By 1926, Enrico Fermi, at the age of twenty-five, was a fully established personality on the

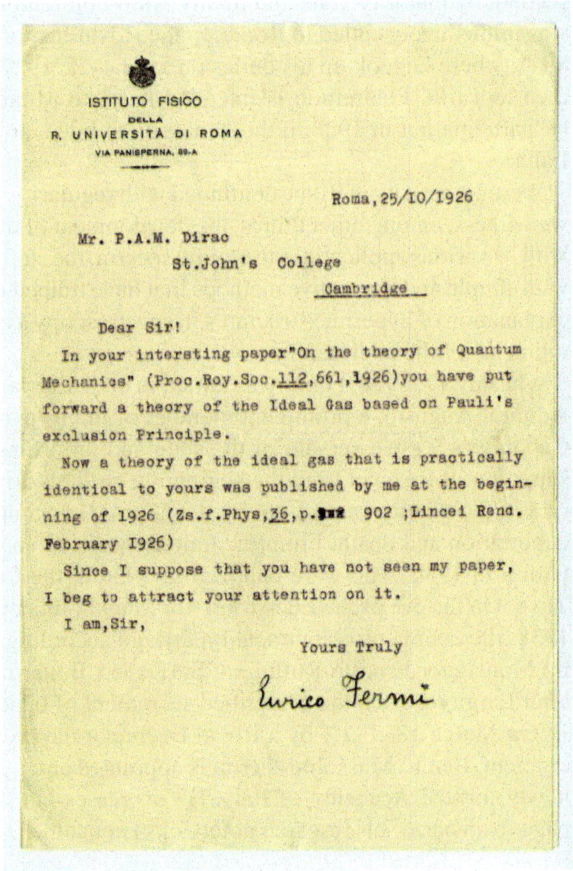

ISTITUTO FISICO
DELLA
R. UNIVERSITÀ DI ROMA
VIA PANISPERNA, 89-A

Roma, 25/I0/I926

Mr. P.A.M. Dirac
St.John's College
Cambridge

Dear Sir!

In your intersting paper"On the theory of Quantum Mechanics" (Proc.Roy.Soc.112,661,1926)you have put forward a theory of the Ideal Gas based on Pauli's exclusion Principle.

Now a theory of the ideal gas that is practically identical to yours was published by me at the beginning of 1926 (Zs.f.Phys,36,p. 902 ;Lincei Rend. February I926)

Since I suppose that you have not seen my paper, I beg to attract your attention on it.

I am,Sir,

Yours Truly

Enrico Fermi

**Fig. 6** Fermi letter to Dirac, October 25, 1926

international scientific scene. His output numbers some fifty papers of highest level, on a wide range of topics including electromagnetism, general relativity, Röntgen rays, mechanics, ergodic theory, the photoelectric effect, the statistical mechanics, quantum mechanics, and spectroscopy. Fermi's scientific output also includes experimental works, such as those pioneering works on the effects of an alternating field over the polarization of resonance light, carried out in 1925 in Florence in collaboration with Franco Rasetti, as a prelude to future collaboration.

## *1.3 Fermi at via Panisperna (Rome)*

Meanwhile, Orso Mario Corbino had succeeded in his far-sighted attempt to establish a Chair of Theoretical Physics in Rome, and to have the corresponding national competition, the first in Italy, announced.

The outcome of the competition, for titles only, had the expected conclusion: the triple of winners included Enrico Fermi in first place, unanimously, Enrico Persico in second, by majority vote, and finally Aldo Pontremoli in third, unanimously. Fermi was immediately called to Rome, to the Royal Institute of Physics at via Panisperna 89 A, where he took up his duties on January 1, 1927. Persico goes to Florence and then to Turin. Pontremoli is intended to go to Milan, but then meets a tragic fate by participating in 1928 in the Umberto Nobile's polar expedition with the airship Italia.

Fermi's activity in Rome continued with regularity. In the years from 1927 to 1933, we witness, among other things, the development of the statistical model of the atom, with its various applications to atomic spectra, the study of quantum electrodynamics, with simple and innovative methods that have trained a generation of researchers, the explanation of hyperfine structures in spectroscopy as a consequence of the magnetic moments of the nucleus.

On July 17, 1928, Fermi was united in civil marriage to Laura Capon, daughter of Admiral Augusto, a prominent member of the Roman Jewish community. Augusto Capon was a great servant of the state and unfortunately an honest and consistent supporter of Fascism from the beginning, until his arrest in October 1943 during the raid in the ghetto, carried out by the Germanic occupation forces, which led to his deportation and death. From the union two children were born, Nella in 1931, and Giulio in 1936, who were baptized in 1936 at the Church of Santa Bonosa on via Tirso. On the eve of their departure for Stockholm, and the United States, on Dec. 5, 1938, the couple also contracted marriage according to the Catholic rite, officiated by Monsignor Ernesto Ruffini, at St. Robert Bellarmino Parish in Piazza Hungary, after Laura received the prescribed sacrament of baptism.

On March 18, 1929, by a Royal Decree at the proposal of the Head of the Government, Benito Mussolini, Fermi is appointed among the first thirty members of the newly formed Academy of Italy. He expresses his satisfaction with the event in a page from one of his research notebooks kept at the Domus in Pisa, writing in capital

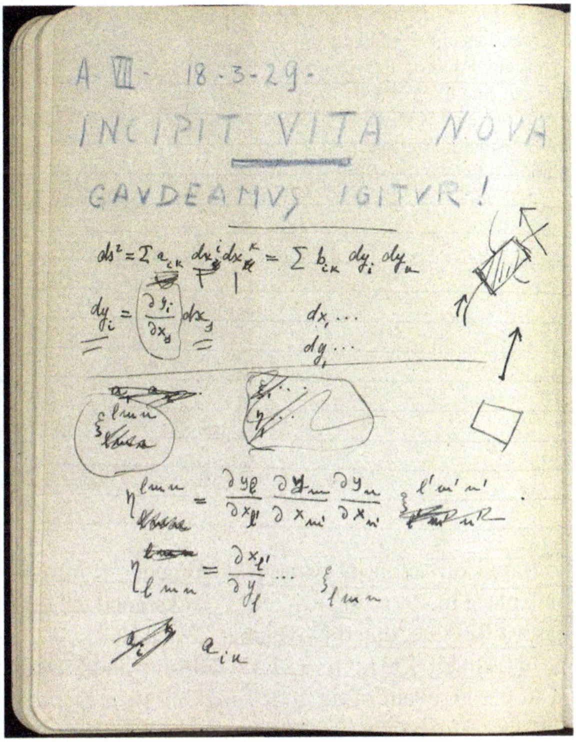


**Fig. 7** Fermi research notebook (1928) (Domus Galilaeana, Pisa)

letters with blue pencil the sentence: "A - VII- 18-3-29 - INCIPIT VITA NOVA - GAUDEAMUS IGITUR."

This is one of the very few personal expressions in a research notebook by Fermi, where he takes up, in the appropriate "academic" context, the words of a goliardic hymn certainly learned during his years of study in Pisa: "Gaudeamus igitur, Iuvenes dum sumus." (See Fig. 7).

An effective cunning of the Regime had been to grant academicians not only symbolic rights of precedence in official ceremonies, immediately after the Deputy Attorney General of the State, but also a kind of congruous prebend of as much as 36,000 Lire annually, corresponding to about twice of a university professor's salary.

The official date of Fermi's membership in the National Fascist Party is that of April 27, 1929, about forty days after his appointment to the Academy, so much so that Mussolini, after the expatriation of Fermi to the United States, hastened to say, believed by nobody, that he had proposed him as Academician of Italy in 1929 without even knowing whether he was a Party member. There is, however, a well-known photograph that shows Fermi exhibiting on the left lapel of his jacket a very recognizable "bug" of the Partito Nazionale Fascista (National Fascist Party) while in a boat on Lake Como in 1927, in the company of Werner Heisenberg and Wolfgang Pauli. (See Fig. 8).

**Fig. 8** Enrico Fermi, Werner Heisenberg and Wolfgang Pauli on the Como Lake during the Volta Conference of September 1927 (Amaldi Archive, Sapienza University, Rome)

Based on some ministerial documents, it must be assumed that the inscription took place in 1926. In any case, Fermi signed the prescribed oath of allegiance to the Fascist Regime, dated November 23, 1931.

In October 1931, Orso Mario Corbino and Enrico Fermi succeeded in organizing in Rome an event of an epoch-making nature: a major international conference on nuclear physics, the first in the world in terms of scale and resonance. It is attended by some fifty official guests, including seven Nobel laureates, and many future honorees. This conference shows that by now the focus of interest of Corbino and Fermi has shifted to nuclear physics, which constitutes the real frontier of research after the substantial exhaustion of atomic and molecular physics, with the advent and development of the new quantum mechanics of Heisenberg, Born, Jordan, Dirac, Schrödinger. The conference is very successful and provides a very effective forum for fruitful debate among leading experts in the field on the current status and possible developments in research on the constitution of atomic nuclei. Only a few months later, in the legendary Cavendish Laboratory directed by Ernest Rutherford in Cambridge, James Chadwick succeeded in demonstrating the existence of a new nuclear constituent, the neutron, a particle of mass comparable to that of the proton but without charge.

The discovery of the neutron produces a real revolution in nuclear physics. In particular, Werner Heisenberg and Ettore Majorana succeed in demonstrating that it is possible to construct realistic models of the atomic nucleus, within the framework of existing quantum mechanics, by assuming protons and neutrons as basic elementary constituents, held together by appropriate exchange forces.

In this context Enrico Fermi fits in with the utmost authority and effectiveness. Neutron discovery indeed opens a period of research that shares almost miraculous aspects, in which Fermi alone within a few months, from December 1933 to October of the following year, succeeds in achieving astounding results of absolute international level. In December 1933, starting with the nuclear model of Heisenberg-Majorana, consisting only of protons and neutrons, Fermi formulated an innovative theory of beta decay, in a quantum field theory framework, according to which the beta electron does not pre-exist in the nucleus, but is created at the moment of decay, together with a new particle introduced by Pauli, the neutrino, as a result of the transformation of a neutron into a proton. This respects the conservation of energy and charge, and one accounts for the continuous beta spectrum, as a consequence of the different distribution of available energy, between beta electron and neutrino, which escapes observation.

Soon thereafter, in March 1934, again ALONE and guided by his theory of beta decay,[1] he demonstrates that it is possible to produce induced radioactivity by bombardment with neutrons. The instrumentation used by Fermi in this discovery is extremely simple but effective. It consists of a neutron source consisting of radon and beryllium powder specially prepared by Giulio Cesare Trabacchi, director of the Physics Laboratory of the Institute of Health, with which to irradiate samples, and a Geiger-Müller (GM) counter, with which to measure any radiation emitted by the samples once irradiated.

By bombarding with its neutron source two samples of aluminum and fluorine, for about two hours and for a few minutes, and measuring their activity after the source has been removed, Fermi discovered that they emitted radiation with an exponential trend, typical of a radioactive decay, i.e., that they were transformed into new radioactive elements (with lifetimes average of 12 min and 10 s, respectively). This will soon prove to be an important discovery, with profound implications for the future of nuclear physics and the fate of humanity. In July 2002, we found in the archives of the Oscar D'Agostino Foundation in Avellino the laboratory notebook used by Fermi to record the results that led him to the discovery of neutron-induced radioactivity. (See Fig. 9).

This notebook is very illuminating for understanding the various stages of the discovery. Among other things, it shows us how Fermi built the first Geiger-Müller counters himself (almost certainly with the help of Bruno Rossi, a great expert in the field and his friend) and how he proceeded in choosing the elements to irradiate: the first element he chose was platinum, the second is aluminum, and with this he immediately discovers induced radioactivity.

The discovery of neutron-induced beta radioactivity concerns at first only two elements, aluminum and fluorine, and is immediately communicated by Fermi in a Letter to La Ricerca Scientifica, dated March 25, 1934 (No. 84th).[2]

Within a few days Fermi was able to test on his own dozens of other elements, as he communicates in a further Letter to La ricerca Scientifica (No. 84b) in footnote 2) and in a letter to Nature, sent on April 10 (No. 93 in [3]).

---

[1] Francesco Guerra, Nadia Robotti, The lost notebook of Enrico Fermi: the true story of the discovery of neutron-induced radioactivity, Società Italiana di Fisica, Springer, Rome, Berlin, 2018.

[2] Enrico Fermi, Note e Memorie (Collected Papers), Volume I, Italia 1921–1938, Volume II, United States, 1939–1954, Accademia Nazionale dei Lincei, Roma—The University of Chicago Press.

[3] Enrico Fermi, Note e Memorie (Collected Papers), Volume I, Italia 1921–1938, Volume II, United States, 1939–1954, Accademia Nazionale dei Lincei, Roma—The University of Chicago Press.

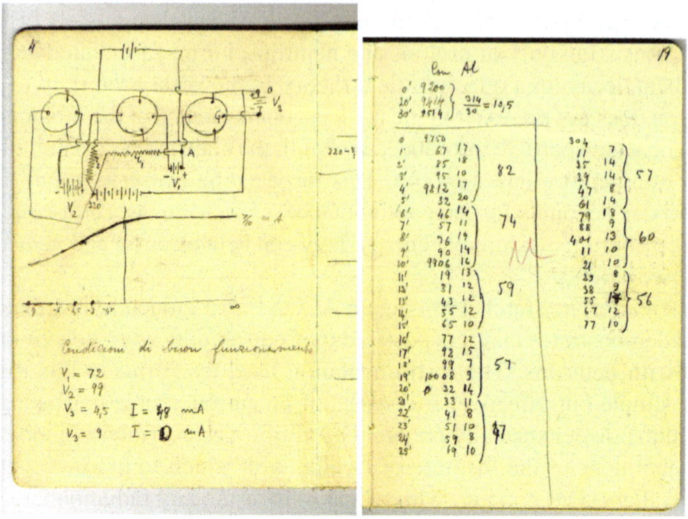

**Fig. 9** Fermi notebook used during his discovery of neutron induced radioactivity (march 1934). On the right, page of Fermi discovery of neutron induced radioactivity in Aluminium

Soon after, Fermi initiated a systematic search on the entire periodic table, drawing on the collaboration of D'Agostino (who, according to the aforementioned laboratory notebook, is his first collaborator in this research), Amaldi, Segrè, Rasetti and finally Bruno Pontecorvo.

Crucial is the constant support offered by Trabacchi in renewing the sources. In a short time Fermi and his group manage to activate about two-thirds of the elements of the periodic table. By bombarding thorium and uranium, some results are found that are difficult to interpretation, which are immediately attributed to the creation of 'transuranic elements.'

It was instead, as would later be understood, fission products. On October 20, 1934, another major discovery was made by Fermi, that of the effect of the neutron slowing down. By interposing a block of paraffine between the source and a sample of silver, Fermi notices that the induced radioactivity is greatly intensified. Effects similar are found for many other elements, and also by using water or other hydrogenated substances.

As Fermi immediately explains, neutrons when they collide with the hydrogen nuclei contained in these substances are slowed down, and slowed ('slow') neutrons have a larger cross section for many nuclei. We now know that slow neutrons are a fundamental key to accessing nuclear energy. This discovery, at Corbino's suggestion, is immediately patented (Oct. 26 October 1934) on behalf of "E. Fermi, E. Amaldi, O. D'Agostino, F. Rasetti, B. Pontecorvo, G.C. Trabacchi, E. Segrè."

In May 1935, for a variety of contingent reasons, the group disbanded, and at the Institute of physics in Rome, Fermi and Amaldi, and occasionally Rasetti and Wick, remain to work on the nuclei.

## *1.4 Towards America*

After the discovery of the effects of slow neutrons, Fermi abandons radiochemical research, aimed at the chemical identification of the radioactive nuclides produced, and devoted himself with foresight, to the systematic study of slowing, scattering, and absorption of neutrons in various substances. Distinguished members of the international scientific community, including Werner Heisenberg, Victor Hess, Max Planck, James Franck, Arthur Compton, Erwin Schrödinger, Chandrasekhar, Raman, proposed Fermi already for the 1935 Nobel Prize, with subsequent re-proposals for 1936, 1937, 1938. Finally, the Swedish Academy of Sciences awarded him the Nobel Prize in Physics 1938, (see Fig. 10) with the following motivation:

> ... for Your discovery of new radioactive substances belonging to the entire field of elements and for the discovery, made by you in the course of such studies, of the selective power of slow neutron.

This is an all-Italian Nobel Prize. Fermi was able to achieve epoch-making results in nuclear physics, making the best exploitation of the scarce resources existing in Italy, helped by his physical intuition, but also by his courage and determination. Well before Enrico Fermi was awarded the Nobel Prize, he realized that expatriation was now inevitable. A complex of very different reasons enter decisively.

Certainly counts the impossibility of continuing research in Italy, due to the lack of adequate instrumental means. In fact, the agile and reliable "homemade" neutron sources, consisting of a glass vial filled with radon and beryllium powder, with which Fermi had made his discoveries, no longer made it possible to pursue research competitively internationally. Much more powerful sources were needed, which only accelerator particle could ensure. But a cyclotron, even of modest power, was far beyond of the concrete possibilities offered by a rather underdeveloped "Italietta", albeit one with imperial vocations.

There are then for Fermi serious problems related to the exploitation of patents on neutrons in Italy, that is, in an environment not very favourable from the economical and industrial point of view. Fermi, in fact not only holds the well-known patent on slow neutrons, together with his collaborators, but also applied, in 1935, for another patent, in his name only, involving all interactions of neutrons with matter. This application, filed in the United States and Canada, has a very troubled life, and is contrasted in every way, even by some of his collaborators in Rome. Thus a total inclusion of Fermi in the United States certainly would have favoured the exploitation of these two patents.

We must also consider that the socio-political context in Italy is undergoing a very dangerous drift. The totalitarian character of the state is being exacerbated. Politics is

**Fig. 10** Photo of the Fermi awarding of Nobel price. Fermi's frack still exists and is on display in Rome, at the Museo Storico della Fisica e Centro Studi e Ricerche Enrico Fermi

closer and closer to that of Nazi Germany. Italy is on its way to enacting laws that are racial laws absolutely unacceptable in their infamy. In addition, the possibilities of an imminent generalized conflict, called "guerrone" in some of the letters of Bruno Pontecorvo. (See Fig. 11).

Fermi's expatriation is carefully planned. First, on September 4, 1938 (the Nobel Prize had not yet been announced) he let it be known to Columbia University that the reasons that had led him, two years earlier, to reject the offer of a permanent faculty position are no longer valid and that he gladly accepts a position at Columbia University, or even elsewhere. The date of the letter is in significant and timely correlation with the measures adopted by the Council of Ministers on September 2 and 3 September, regarding the exclusion of Jews from teaching at every grade and order, and the prohibition of pupils of the "Jewish race" from attending public schools.

Moreover, on September 3 itself, the Accademia d'Italia, of which Fermi is a member, formed a committee of academicians with the task of studying "what were through the centuries the manifestations and reflections of Judaism in the life of Italy from the time of ancient Rome to the present."

**Fig. 11** Bruno Pontecorvo with Enrico Fermi. (Visit to Olivetti Factory in Ivrea, Italy, 1949)

Fermi also takes advantage of all legal possibilities to be able to expatriate with the full consent of the Italian authorities. Upon receiving the news of the award of the Nobel Prize, he obtains permission to attend the Prize award ceremony in Stockholm, and then travel to the United States, on extraordinary leave, to deliver a lecture series at Columbia University.

At the end of the period of leave, he obtains a leave of absence for family reasons, without allowance. At the end of the period of leave, in 1940, he informed the Ministry that he could not return to take up service, because he is involved in very important research. The Ministry accepts the resignation. On his chair of Theoretical Physics, at the end of 1940, Gian Carlo Wick, the leading theoretical physicist left in Italy, was eventually called.

Then, after collecting the Diploma and Medal of the Nobel Prize in Physics from the hands of His Majesty King Gustav V of Sweden, in the sumptuous ceremony at the Concert Hall in Stockholm on December 10, 1938, Enrico Fermi with his family traveled to Copenhagen for a very brief stay with Niels Bohr, and then embarked with his family directly to the United States, escaping the backward and oppressive atmosphere of the Fascist Regime, and the horror of the racial laws and the impending conflict in Europe.

## 1.5 Fermi in America

By a singular coincidence, Fermi's emigration to the United States takes place at the same time as Otto Hahn's discovers in Berlin a new physical phenomenon, the fission of the uranium nucleus by bombardment with slow neutrons, capable of releasing enormous amounts of energy.

Fermi, first at Columbia University and then at the University of Chicago, embarked on the enterprise of harnessing the energy released by fission for practical purposes, through a controlled chain reaction. This is a problem of scientific and technological formidable complexity. Indeed, the uranium nucleus that undergoes fission by slow neutrons is the rare isotope 235, present only with 0.7 percent fraction in natural uranium. In the course of fission are released fast neutrons, which will have to be slowed down, via a "moderator."

In a systematic way Fermi is involved first of all in determining the number of neutrons liberated for each fission, in choosing the moderator kind, in determining the best geometry on how to arrange uranium and moderator, so that the moderator slows down the neutrons without absorbing too many, keeping in mind that the most abundant isotope in natural uranium, 238, is an avid neutron absorber at a resonance energy that lies between that of the neutrons produced and that of the thermal neutrons useful to fission. With this research, Fermi is a full participant in the Manhattan project, aimed at building nuclear weapons, to be used possibly in the ongoing war.

It is compelling to follow in detail, through the study of their respective "secret reports", the long-distance scientific confrontation between Enrico Fermi and his current formidable rival, Werner Heisenberg, involved in the similar German nuclear project. Significantly, both developed independently the theory of the nuclear reactor, in a physically equivalent manner, building on Fermi's work on the slowing down, scattering and absorption of neutrons in matter, carried out in 1935–1938, before the discovery of fission. But let us follow some key passages of the whole story.

Thus we learn that still in March 1943 Fermi was discussing with Harold Urey on the possibility of using heavy water as a moderator, as described in the secret report A-544 (No. 184 in Footnote 2). While Heisenberg, through a beautiful experiment [4] to evaluate the diffusion length of thermal neutron in heavy water, performed in Leipzig in collaboration with Robert and Klara Döpel, comes as early as August 1940 to the clear conclusion that "aus D2O und 238 eine Maschine zur Energieerzeugung gebaut werden kann" ("it will be possible to construct a device for the exploitation of energy between heavy water and natural uranium").

On the other hand, Fermi determines as early as September 25, 1940 the correct absorption cross section of diffusion neutrons on carbon (Secret Report A-21, No. 136 in[4]), while Heisenberg must refer to an elegant experiment by Walther Bothe and Peter Jensen of January 1941, described in secret report G-71.

Unfortunately, Bothe and Jensen's experiment comes to the fatal conclusion that "The gemessen Diffusions länge ist soviel kleiner, die Absorptionalso soviel stärker,

[4] Werner Heisenberg, Gesammelte Werke, Collected Works, Springer-Verlag 1984.

dass die hier untersucht Kohle als Bremssubstanz für die Maschine kaum in Betracht kommen dürfte" (The measured length of diffusion is so much smaller, and the absorption is also so much stronger that carbon examined here cannot be considered as a convenient slowing substance for the machine). One cannot doubt the skilled experimental ability of Bothe and Jensen, and the validity of their apparatus, which moreover after all, derives from well-known methods developed by Fermi as early as 1936. The Siemens electro-graphite, available to them at that time, was loaded with small, strongly neutron-absorbing impurities, for which it was difficult to make the proper correction.

What is certain is that Heisenberg in 1940–1941 found himself automatically steered in the direction of choosing heavy water as moderator. A choice that is scientifically correct, which leads him, however, to run up against the chronic scarcity in supplies of pure heavy water and natural uranium, in the Germania at war at the time. In the last B8 experiment carried out in Heigerloch early 1945, with Allied troops approaching, the reactor of Heisenberg produces a strong overall multiplication in the number of neutrons, but does not reach the critical condition.

Instead, Fermi about the same time decides on graphite, which the American industry can supply him with at the highest purity. It is well known that on 2 December 1942 Fermi succeeds in triggering the self-sustained controlled chain reaction in an experimental reactor using natural uranium and graphite, placed under the bleachers of a decommissioned stadium at the University of Chicago.

This is a key strategic achievement, leading to the construction of power reactors, capable of producing plutonium by neutronic irradiation of uranium rods. Plutonium is easily separated from uranium by chemical route, and is a highly fissile substance. As part of the Manhattan Project, the Trinity Test of July 1945, shows the effectiveness of plutonium as a nuclear explosive.

By decision of the U.S. American administration, two nuclear bombs are dropped on Japan at Hiroshima and Nagasaki. The first is based on uranium-235, available following the success of the natural uranium enrichment program, the second is based on plutonium produced in power reactors.

In the postwar period, Fermi is at the acme of his success. By now, since 1944, he is an American citizen. Among the many awards, he is awarded by Congress in 1946 of the Medal of Merit, the highest civilian decoration in the United States. (See Fig. 12).

He can continue his research having at his disposal impressive means. Consider that the neutron flux from a nuclear reactor can reach to the fantastic value of one million billion per square centimeter per second, to be compared with the value of one hundred thousand per second of its first radon-beryllium sources. With these new sources Fermi can perform elegant neutron diffraction experiments on nuclei. Fermi also fits effectively into the development and applications of the early electronic computers, arriving, for example, at the evaluation of the energy distribution in nonlinear dynamical systems (Fermi-Pasta-Ulam), a research direction directly related to his early work in statistical mechanics.

**Fig. 12** Attestation and Medal for Merit assigned by United States of America to "Dr Enrico Fermi," January 12, 1946. All the medals Fermi received, including the Nobel Prize medal, are kept at the Chicago Library in Chicago

Fermi also proves to be a pioneer in high-energy physics, studying the scattering of pions, produced at the Chicago cyclotron, on protons, and arriving at the discovery of pion-nucleon resonances, among the first new artificially created unstable particles.

Also of notable importance is the development of a simple model, essentially valid to this day, which makes it possible to explain the very large energy of cosmic rays as due to collisions, mediated by magnetic fields, of nuclei nuclei, particularly protons, with astronomical-sized structures. In other words, cosmic rays would participate in a kind of equipartition of energy with the much larger bodies in the Universe. Unfortunately, an incurable disease cut him short on November 28, 1954, at only

53 years old, ineluctably interrupting the further series of scientific achievements he could have achieved.

After the War, Fermi returned to Italy two times, in 1949 and 1955. During the first visit in 1949, Fermi was invited by the Academia dei Lincei and the Foundation "Guido Donegani" to give conferences on subjects of atomic Physics of current high interest. Six conferences were held in Rome, at the Institute of Physics in the University, and three in Milan, in the great Hall of the Montecatini Society. Each conference was followed by a young professor, who took care of the publication of the text in a volume edited by Accademia dei Lincei. Involved people included Sebastiano Sciuti, Lucio Mezzetti, Ettore Pancini, Nestore Bernardo Cacciapuoti, Mario Ageno, Giacomo Morpurgo, Carlo Salvetti (two conferences), Piero Caldirola. Their valuable contributions are offered to the much larger community of english speaking people, in this volume.

# Enrico Fermi and the Communication of Physics

**Vincenzo Barone**

## 1 Spreading the New Physics

In January 1922, a few months before graduating from the University of Pisa and the Scuola Normale Superiore, Enrico Fermi described his activities to his friend Enrico Persico as follows:

> I am acting as lecturer, relativist, and physicist. In the first activity, as you probably have heard, I have not yet sullied my conscience with manslaughter. The philosophers have become terribly angry with me because, in their own words, "since the bases of the theory of relativity are not only physical but also logical, it would be advisable to put oneself in touch also with the results which this logic has attained". But Wer "fregiert" sich darum:[1] until they hit me they don't scare me.[2]

In the 1920s, after the verification of the gravitational deflection of light and the explosion of Einstein's planetary fame, relativity had become a hot topic of debate, beyond the scientific boundaries. Even in Italy, many philosophers began to discuss

[1] Who cares?

[2] E. Fermi to E. Persico, January 25, 1922, in E. Segré, Enrico Fermi, Physicist, University of Chicago Press, Chicago 1970, p. 197.

V. Barone (✉)
DiSIT, Università del Piemonte Orientale, Alessandria, Italy
e-mail: vincenzo.barone@uniupo.it

INFN, Sezione di Torino, Turin, Italy



A. Treves and P. Tucci (eds.), *Enrico Fermi, Atomic Physics Lectures*, History of Physics,
https://doi.org/10.1007/978-3-031-68690-0_2

it, without bothering to really grasp it.[3] It was with some of these scholars that Fermi probably found himself confronted, as he told Persico. In the same letter he added that he was uncertain about the publication of one of his public lectures on relativity in "Scientia" (a journal of scientific synthesis), because—he said—"I am little afraid that the philosophers might rebel in fury and draw me into fruitless polemics".

The paper for "Scientia" never saw the light of day, but relativity was the subject of Fermi's first popular essay, a short contribution to the Italian edition of August Kopff's book on Einstein's theory,[4] appeared in 1923. The editors of the volume had decided to add a long appendix, entitled *Valore e interpretazione della teoria della relatività* [Value and Interpretation of the Theory of Relativity], containing observations and comments from leading Italian scientists and philosophers. These writings clearly show how the Italian physics and astronomical community still maintained an attitude of hostility and incomprehension towards relativistic theories. The only exception is the contribution by Fermi, then 21 years old[5] and virtually unknown. *Le masse nella teoria della relatività* [The masses in the theory of relativity][6] is a jewel of scientific prescience, truly surprising for such a young author. While the attention of physicists was still mainly focused on the optical consequences of relativity,[7] Fermi pointed to the relativistic equivalence of mass and energy as the most important and fruitful aspect of Einstein's theory, often obscured by the debates on the alleged space-time paradoxes:

> The great conceptual importance of the theory of relativity, as a contribution to a deeper understanding of the relations between space and time, and the lively and often passionate discussions to which it has consequently given rise, even outside strictly scientific circles, have perhaps somewhat distracted the attention from another of its results. This, though less clamorous and, let us admit it, less paradoxical, nevertheless has consequences in physics that are no less noteworthy, and whose interest is likely to grow in the future development of science. The result to which we refer is the discovery of the relationship that links the mass of a body to its energy. The mass of a body, according to the theory, equals its total energy divided by the square of the speed of light. Even a superficial examination shows us that, at least for physics as it is done in laboratories, the importance of this relationship

---

[3] See: R. Maiocchi, Einstein in Italia. La scienza e la filosofia italiane di fronte alla teoria della relatività, Franco Angeli, Milano 1985. S. Linguerri and R. Simili (eds.), Einstein parla italiano. Itinerari e polemiche, Pendragon, Bologna 2008.

[4] A. Kopff, I fondamenti della relatività Einsteiniana, R. Contu and T. Bembo eds., Hoepli, Milano 1923.

[5] The draft of Le masse nella teoria della relatività is dated "October 1922", three months after Fermi's Laurea at the Scuola Normale. The date was omitted in the printed text. The galley proofs of this three-page essay, signed "Dott. E. Fermi", were recently sold at auction in United States (https://www.rrauction.com/auctions/lot-detail/345300506220134-enrico-fermi-signed-galley-proof).

[6] E. Fermi, Le masse nella teoria della relatività, in A. Kopff, I fondamenti della relatività, Ref. 4, pp. 342–344 (FNM, 5). Fermi's papers are collected in E. Fermi, Note e Memorie, Accademia Nazionale dei Lincei and University of Chicago Press, vol. I, Roma 1961; vol. II, Roma 1965. From now on, these volumes will be referred to as FNM, followed by the paper number.

[7] M. De Maria and G. Maltese, I fisici sperimentali italiani e la relatività (1906–1927), "Quaderni di Storia della Fisica" 1 (1997), pp. 245–273. G. Maltese, The late entrance of relativity into Italian scientific community, "Historical Studies in the Physical and Biological Sciences", 31 (2000), pp. 125–173.

between mass and energy is such that it considerably obscures that of the other consequences, quantitatively very small, but to which our mind gets used with more effort.

The relationship between mass and energy (the famous $E = mc^2$)—explains Fermi—"certainly leads to huge numbers":

> For instance, if it were possible to release the energy contained in one gram of matter, one would obtain an energy greater than that developed in three years of uninterrupted functioning by a thousand-horsepower engine (no need for comments!). One could rightly say that it does not seem possible that, at least in the near future, a way will be found to release this frightening energy, which is only to be hoped for, because the explosion of such a terrible amount of energy would first of all break into pieces the physicist who had the misfortune to find a way to produce it.

"But even if such a complete explosion of matter does not appear to be possible for the moment"—Fermi continues—"experiments aimed at achieving the transformation of chemical elements into one another have been underway for some years" and these processes "are linked to energy exchanges that the relationship between mass and energy allows us to study very clearly". The numerical estimates "show us that the energy of nuclear bonds is a few million times greater than that of the most energetic chemical bonds and explain why the problem of the transformation of matter, the dream of the alchemists, has been the subject for many centuries of the unsuccessful efforts by the cleverest minds, and how it is only now, using the most energetic means at our disposal, that we have succeeded in obtaining this transformation, but in quantities so small that they escape the most delicate analysis". In conclusion, says Fermi, "the theory of relativity, not only gives us a clear interpretation of the relations between space and time, but will, perhaps in the near future, be destined to be the keystone for solving the problem of the structure of matter, the ultimate and most difficult problem of physics".

It makes a certain impression to think that it is the young author of this paper who will achieve, many years later, the sensational results foreshadowed in it.[8] But Fermi's juvenile essay is also valuable from another point of view, as a masterful example of popular science writing. Fermi, in fact, had a great ability to capture the reader's attention through examples of considerable impact, and to go back—in an equally compelling manner—to the terrain of real science, with a balanced dosage of suggestions and scientific data.

Today we remember Fermi as a hard and pure scientist, totally immersed in his studies and uninterested in everything else. However, since his debut in research, he also engaged in an intense work of popularization. Finding himself—almost alone in Italy—on the most advanced front of the new physics, he felt it was his task to disseminate the discoveries that were accumulating in the study of the fundamental phenomena of nature. And he did this with great effectiveness, exploiting his natural predisposition to make even the most complicated topics fully understandable. In his famous biography of his friend and teacher, Emilio Segrè recalls the first time he met him, at the end of 1923:

> I had been told by a schoolmate, Giovanni Enriques (son of the mathematician Federigo) that an extraordinary young man named Fermi was to lecture on quantum theory, and I went to

---

[8] Segrè speaks, in this regard, of an "almost prophetic premonition" (FNM, vol. I, p. 33).

the lecture. [...] I was very much impressed by the clarity of the presentation, the significance of the discussions, and the speaker's knowledge and self-assurance.[9]

The lecture that impressed Segrè so much, entitled *Sui principii della teoria dei quanti*[10] [On the Principles of Quantum Theory], was delivered by Fermi on 24 November 1923 at the Mathematical Seminar of the University of Rome. Three years later, at only 25 years of age, he was appointed by the same University to the first Italian chair of Theoretical Physics. The establishment of this chair, strongly supported by Orso Mario Corbino, director of the Physics Institute, and by the mathematical group (Tullio Levi-Civita, Federigo Enriques, Guido Castelnuovo), had a twofold purpose: on the one hand, to guarantee a prestigious position to the man who was considered Italy's greatest scientific promise, while at the same time offering him full political, financial and academic support for the establishment of a research group that would bring Italy up to the level of the most advanced nations; on the other, to officially launch theoretical physics as an autonomous field of research and teaching in the Italian academic system.

Fermi's appointment at Rome marked the beginning not only of a fruitful season of theoretical and experimental research, but also of an intense campaign to spread the new physics in all directions: among colleagues, school teachers, young people and the educated public.[11] In this activity Fermi was joined by his friend Enrico Persico, professor of theoretical physics in Florence and then in Turin. For many years, the two alternated at meetings and on the pages of cultural journals, dividing their tasks according to their respective inclinations: Persico more attentive to conceptual implications, Fermi more adherent to strictly scientific aspects.

Even before becoming a professor, Fermi began to collaborate with the Mathematical Seminar of the University of Rome and with the "Periodico di Matematiche" (the journal of "Mathesis", the association of secondary school teachers of mathematics and physics), both directed by Federigo Enriques. In the period 1923–1930, Fermi gave several lectures at Enriques' Seminar, after the one attended by Segrè: the others concerned atomic physics, solid state physics and quantum mechanics. From 1926 onwards, he also assiduously attended the annual meetings of the SIPS, the Italian Society for the Advancement of Science, which provided an important opportunity for the communication of scientific results and interdisciplinary debates. At the SIPS meetings, he gave a number of technical talks as well as general lectures on atomic and nuclear physics. His propaganda for the new physics also took the form of essays published in cultural and popular magazines, of various articles in the Enciclopedia Italiana (of which he was first a collaborator, and from 1932

[9] E. Segrè, Ref. 2, p. 24. See also E. Segrè, A Mind Always in Motion, University of California Press, Berkeley-Los Angeles 1993, p. 41.

[10] E. Fermi, Sui principii della teoria dei quanti, "Rendiconti del Seminario matematico dell'Università di Roma", serie II, 2 (1925), pp. 7–12 (FNM, 22).

[11] See: C. Petrucci and F. Sebastiani, L'opera di divulgazione della fisica quantistica svolta in Italia da Enrico Fermi negli anni Venti, "Quaderni di Storia della Fisica", 14 (2007), pp. 49–70; V. Barone, Il sollievo della semplicità. Enrico Fermi fisico e divulgatore, in E. Fermi, Atomi Nuclei Particelle. Scritti divulgativi ed espositivi 1923–1952, V. Barone ed., Bollati Boringhieri, Torino 2009.

director of the physics section[12]), and of a treatise—the first in Italy—on quantum physics.[13] Finally, it should not be forgotten that in 1934—in the midst of his experimental studies on neutrons—Fermi made a long trip to South America, during which he gave a dozen public lectures, on topics ranging from atomic theory to the structure of nuclei, from quantum mechanics to artificial radioactivity.[14]

It was therefore a wide-ranging pedagogical and cultural program that Fermi implemented in the decade 1925–1935, which was also his most intense period of research, demonstrating that the two activities—science and its communication—were systematically intertwined, without one being to the detriment of the other.

A similar program was carried out by Fermi in the post-war period, with the aim, on the one hand, of decoupling the discourse on atomic energy from the question of the bomb, emphasizing the peaceful uses of nuclear physics,[15] and, on the other hand, of drawing everyone's attention back to pure research. In a speech at Union College in 1947, Fermi spoke of a "crisis" in science brought about by its involvement in wartime events: "The profession of the man of research—he said—must go back to its tradition of research for the sake of uncovering new truths. Because in all directions we are surrounded by the unknown and the vocation of the scientist is to push back the frontiers of our knowledge in all directions, not only in those that show promise of more immediate gains or more immediate applause".[16] Fermi spoke about fundamental science in the two most important lecture series of that period

---

[12] The main entries of the Enciclopedia Italiana written by Fermi are: Teoria elettrica dell'atomo (Vol. V, 1930), Elettrone (Vol. XIII, 1932), Statistica, Meccanica (Vol. XXXII, 1936), Atomo (I Appendice, 1937). See S. D'Agostino and A. Rossi (eds.), Enrico Fermi e l'Enciclopedia Italiana, Istituto della Enciclopedia Italiana, Roma 2001.

[13] E. Fermi, Introduzione alla fisica atomica, Zanichelli, Bologna 1928. This textbook, still based on the old quantum theory was soon replaced by Persico's textbooks, Lezioni di meccanica ondulatoria (CEDAM, Padova 1930, 1935) and Fondamenti della meccanica atomica (Zanichelli, Bologna 1939). Fermi also wrote a school textbook, Fisica ad uso dei Licei (Zanichelli, Bologna 1929). See L. Malagoli, Enrico Fermi come insegnante e divulgatore, in "Atti del XVIII Congresso nazionale di storia della fisica e dell'astronomia", edited by P. Tucci, Como 1998, http://www.sisfa.org/wp-content/uploads/2013/06/xviiiMalagoli.pdf.

[14] A. De Angelis and J.M. Kenny, Enrico Fermi in Argentina and his Lectures in Buenos Aires, Córdoba and La Plata, "Quaderni di Storia della Fisica", 24 (2020), pp. 117–153. F. Caruso e A.J. Marques, Viaggio di Enrico Fermi in Sudamerica in 1934, "Quaderni di Storia della Fisica" 25 (2021), pp. 3–15. I. de Castro Moreira, A viagem de Fermi ao Brasil em 1934 e suas entrevistas e conferências, "Revista Brasileira de Ensino de Física", 44 (2022), e20220204.

[15] See, e.g., E. Fermi, Atomic Energy for Power, The George Westinghouse Centennial Forum, May 1946 (FNM, 224).

[16] Quoted in G. Maltese, Enrico Fermi and the Birth of High-Energy Physics after World War II, in Proceedings of the International Conference "Enrico Fermi and the Universe of Physics" (Rome, 2001), ENEA-SIPS, Roma 2003, p. 234. Fermi pronounced the same words in a speech at the Haverford College in 1953: Enrico Fermi Collection, SCRC, University of Chicago Library, Box 53, Folder 13.

(addressed to students and a general audience): the Donegani Lectures on Atomic Physics in Rome and Milan in October 1949 (translated in this volume) and the Silliman Lectures[17] on "The Growth of Physics in the Past Twenty Years" at Yale University in April 1950.

## 2 A Simplifying Mind

All testimonies about Fermi converge on the fact that his fundamental gift, as a scientist, as a teacher and as a communicator, was simplicity. Persico speaks of Fermi's innate love of simplicity, which manifested itself as much in his scientific and teaching work as in his lifestyle. "If I were to sum up in a single sentence Enrico Fermi's very complex mental physiognomy,—he says—I would say that his prodigious capacity of immediately spotting the essential element in everything and getting to the heart of it in the simplest way was his principle natural gift".[18] Simplicity was for Fermi both a method of research and a style of communication: the affinity between these two aspects is one of the most visible and distinctive features of his work. The path of discovery and the exposition of a problem obeyed similar rules and mechanisms, and even coincided. Fermi's theoretical elaborations often took place in public, at the very moment when he illustrated a certain topic to his students and collaborators, reasoning out loud, at a constant pace—as all those who were fortunate enough to hear him remember.[19] In this way, for instance, his quantum theory of radiation[20] was born, out of an effort to understand and put into a more familiar and simpler form some earlier attempts at quantizing electrodynamics, and out of a continuous effort to refine the presentation of the results.

Fermi's teaching style is described as follows by Edoardo Amaldi:

> He managed to simplify things a lot. Indeed, someone later accused him (not us, who were very satisfied) of oversimplifying, presenting things as easier than they were. And in a way that was true. The fact is that Fermi thought that one had to get to understand the problem right away, and that therefore there was nothing wrong in skipping certain steps so as not to get lost in the difficulties. When one had understood the problem as a whole, he would then begin to explain that things were not so simple: "I have not told you everything: in fact I have skipped this point and that point, and here there is a difficulty that needs to be

[17] The Silliman Lectures covered the following topics: quantum mechanics, elementary particles, the origin of chemical elements, and the nature and origin of cosmic radiation (Enrico Fermi Collection, SCRC, University of Chicago Library, Box 46, Folder 3). On the same occasion Fermi gave some lectures to physics students on elementary particles, later published in the book Elementary Particles, Yale University Press, New Haven 1951.

[18] E. Persico, Souvenir de Enrico Fermi, "Scientia", 49 (1955), pp. 316–324; engl. transl., Commemoration of Enrico Fermi, in Enrico Fermi: His Work and Legacy, C. Bernardini and L. Bonolis eds., SIF-Springer, Bologna-Berlin 2004, pp. 36–44 (p. 42).

[19] See F. Rasetti, Enrico Fermi e la fisica italiana, Celebrazioni Lincee 12, Accademia Nazionale dei Lincei, Roma 1968, p. 11; E. Segrè, Ref. 2, pp. 54–56.

[20] E. Fermi, Quantum Theory of Radiation, "Reviews of Modern Physics", 4 (1932), pp. 87–132 (FNM, 67).

deepened". This fact of showing the essential line at once, perhaps on a path that was not quite correct (or rather it was correct, but there were passages that were not clarified), was a great technique.[21]

The linearity of Fermi's arguments and the fluidity of his speech could give the impression of effortlessness. But, as one of his famous students in the American years, Chen Ning Yang, explains, this impression is deceptive: "The simplicity was the result of careful preparation and of deliberate weighing of different alternatives of presentation".[22]

Fermi, as we have said, covered all the domains of scientific communication: research articles, textbooks, more general papers (mainly intended for colleagues from other disciplines and teachers), popular works in the strict sense (i.e. addressed to a wide audience). Usually, in these four types of texts, authors adopt very different modes of presentation. For example, analogical discourse, iconic representations and qualitative reasoning, which are very common in works aimed at the general public, are rare or almost absent in the more technical documents.[23] In Fermi's case, on the contrary, what is surprising is the formal and stylistic homogeneity of his writings. Whether it is a work of research or popularization, he always adopts the same "communication strategy",[24] based on certain constant elements: (i) a self-consistent presentation, which develops the subject from the beginning, providing the reader with all the necessary information; (ii) some starting points taken for granted (mainly experimental results); (iii) a precise enunciation of the problems, in which they are reduced to the essentials and stripped of all complications, and possibly reformulated in clearer and more elementary terms; (iv) a gradual approach to the solution, guided by physical intuition; (v) an absolute economy of technical means and mathematical formalisms; (vi) an extensive use of concrete examples, models and analogies; (vii) a certain caution with regard to novelties not yet consolidated.

This strategy is clearly visible in the Donegani Lectures, both in the popular and in the more technical ones. See, for example, the way in which Fermi deals with the complicated question of quantum electrodynamics in the sixth lecture. The scheme is the one described above: the problem (the appearance of infinities in the calculation of the interaction between an electron and the electromagnetic field); the experimental evidence (the Lamb shift of hydrogen); the approximation (Bethe's non-relativistic approach); the theoretical refinement (Tomonaga's equation and Schwinger's renormalization procedure); the caution ("It is probably not yet the ultimate theory, but it has brought us closer and closer to understanding the phenomena. [...] Almost all young theoretical physicists are eager to deal with field theories, perhaps even

---

[21] E. Amaldi, Intervista sulla materia dal nucleo alle galassie, edited by P. Angela, Laterza, Roma-Bari 1980, p. 11. On Fermi as a teacher see the contributions to the workshop Fermi maestro e didatta, "Il Nuovo Saggiatore", 17, n. 5-6 (2001), pp. 5–47.

[22] C. N. Yang, in FNM, vol. II, p. 673.

[23] See, e.g., M. Cloître and T. Shinn, Expository Practice, in T. Shinn and R. Whitley (eds.), Expository Science: Forms and Functions of Popularisation, Reidel, Dordrecht 1985, pp. 31–60.

[24] S. D'Agostino e A. Rossi, Ref. 14, p. XX.

a little too one-sidedly: but only the future will tell" [6][25]). Even what is left out in the lecture—Feynman's unconventional approach—signals a typical attitude of Fermi's: his "rational conservatism", i.e. his tendency to always adopt the most traditional approaches, exploiting them to their extreme consequences, before taking new "revolutionary" paths.

Fermi's pedagogical skills were combined with an exemplary clarity of language, which sometimes came at the expense of stylistic elegance (which he never cared for: he generally wrote on the spur of the moment, almost without corrections). There is an episode that illustrates very well the importance Fermi attached to the transparency and accessibility of texts. Bruno Pontecorvo recalls that, having submitted an article to him for the Enciclopedia Italiana, he, after reading the beginning, sharply told him that he did not understand anything about it, and explained to him how it had to be written: "At least the first part (let's say ten per cent or the first two sentences, if the article is short) must be comprehensible to any educated person".[26] Fermi's writings universally respect this rule, as can clearly be seen in the Donegani Lectures. And even in the case of his scientific papers, one does not have to be a physicist to appreciate and understand the first few pages and the essential sense of the work.[27] A splendid example is given by the famous paper on β-decay (Fermi's theoretical masterpiece), the introduction to which is largely understandable even to non-specialists.[28]

Fermi's favorite heuristic—and communication—tool was analogy.[29] "The most natural hypothesis that one is induced to make when confronted with some unknown thing—he wrote in 1929—is that its behavior is analogous to that of similar things better known".[30] And that this was his usual way of proceeding is confirmed by his wife Laura, who explains how, in exploring the unknown, Fermi's mind worked "by comparisons and analogies".[31]

One of Fermi's happiest analogies is the one underlying his theory of β-decay.[32] The interpretation of this process presented two problems. The first, related to the apparent violation of the law of conservation of energy, could be solved by postulating—as proposed by Wolfgang Pauli—that a neutral particle of very small mass, the neutrino, was produced in the process. The second problem was related

---

[25] The numbers in square brackets refer to the nine Donegani Lectures.

[26] B. Pontecorvo, Enrico Fermi, Studio Tesi, Pordenone 1993, p. 40.

[27] Pontecorvo adds that, according to Fermi, the physical essence of any problem could be explained without the aid of formulas (Ref. 26, p. 52).

[28] E. Fermi, Tentativo di una teoria dei raggi β, "Nuovo Cimento", 11 (1934), pp. 1–19 (FNM, 80a). For an English translation see: F.L. Wilson, Fermi's Theory of Beta Decay, "American Journal of Physics", 36 (1968), pp. 1150–1160.

[29] Gerald Holton defines Fermi "a grand master among analogists". See G. Holton, On the Art of Scientific Imagination, "Daedalus" (Spring 1996), pp. 183–208 (p. 193).

[30] E. Fermi, I fondamenti sperimentali della nuova meccanica atomica, "Periodico di matematiche", 10 (1930), pp. 71–84 (p. 72).

[31] L. Fermi, Atoms in the Family. My Life with Enrico Fermi, University of Chicago Press, Chicago 1954, p. 219.

[32] On this analogy see J.R. Oppenheimer, Analogy in Science, "The Centennial Review of Arts and Science", 2 (1958), pp. 351–373.

to the emission of electrons (β-rays). The most obvious hypothesis was that they pre-existed in the nucleus, but there was no way to construct a quantum and relativistic theory of nuclear electrons. Faced with this difficulty, Fermi suggested that in β decay the electrons were not extracted from the nucleus, but created (together with neutrinos) at the moment the process took place. He was guided by the analogy with the emission of light quanta from atoms: a photon emitted by the atom does not exist beforehand, but is generated at the moment of emission. In the first public presentation of this theory—in Buenos Aires in 1934—he said:

> Assuming the creation of another additional particle is not an entirely new hypothesis, since there is already an example in radiation theory: when an excited atom makes a transition into a lower energy state emitting radiation, a quantum of light, which can be considered as being created upon emission and not as pre-existing inside the excited atom, leaves it. One can thus try to construct a theory of β rays analogous to the theory of irradiation of light quanta.[33]

By the 1940s, various elementary processes were known in which previously nonexistent particles suddenly appeared, and in the second Donegani lecture Fermi used the analogy with radiative processes (and with his own β-decay theory) to formally describe and analyze the new reactions. It should be also noted that the fifth lecture in the series, on nuclear orbitals, is essentially a lengthy discussion of the similarities and differences between the dynamics of atoms and that of nuclei, and the title of the eighth lecture, *Analogie ottiche nelle proprietà dei neutroni* [Optical Analogies in Neutron Properties], speaks for itself.

## 3 Moving Frontiers

Fermi's physics almost always proceeded in an intensive direction,[34] investigating the deepest levels of matter: from atoms to nuclei, and then from nuclei to particles.[35] In 1930 he announced in "Nuova Antologia", a prestigious cultural journal, the transition from atomic to nuclear physics:

> And now that the study of the atom and the molecule is well on its way to being resolved, the attention of physicists begins to turn, with increasing insistence, to a new problem: that of the structure of the nucleus. The nucleus [...] is about ten thousand times smaller than the atom; moreover, its elements are glued together with such great forces that relatively enormous amounts of energy must be directed to it in order to break it, so that so far only a few isolated cases are known in which it has been possible to artificially produce a disintegration of the nucleus. It is true that nature itself, with radioactive phenomena, has given us an example of spontaneous disintegration of certain nuclei. However, this is a phenomenon that, for the

---

[33] E. Fermi, quoted in A. De Angelis and J.M. Kenny, Ref. 14, p. 22.

[34] I adopt the terminology of Weisskopf, who distinguishes between "intensive" and "extensive" research in physics. See V. Weisskopf, The connection between physics and other branches of science, "Supplemento al Nuovo Cimento", 4 (1966), pp. 465–491.

[35] This reductionist tendency did not prevent Fermi from making significant contributions to some fields—statistical mechanics, ergodic theory, fluid dynamics, non-linear phenomena—which are now associated with the "physics of complexity".

time being, we can only observe, without managing to speed it up or slow it down, or any way to influence it.[36]

Anticipating an easy objection to this type of research with no immediate applications, he added:

The impression must not [...] arise in public opinion that the work of scientists is now moving away from life and becoming confused in the pursuit of abstruse and purely abstract ideas. It is true that often their work is not directly determined by a desire to make a contribution to a technical question, but rather by an intimate need to push further and further the boundaries that limit our knowledge. Sooner or later, however, other men of more practical ingenuity will show up who will find ways to use their new knowledge for the further development of civilization.[37]

In October 1931, the first international conference on nuclear physics was held in Rome on the initiative of Fermi and Corbino. It was an event of great scientific importance, but also of considerable political resonance, inaugurated in the presence of Mussolini. Fermi reported on it in an article for the journal "Gerarchia" (directed by Mussolini himself), written perhaps at the suggestion of Corbino (who in the same issue signed a commemoration of Edison), and in any case responding to obvious reasons of political opportunity, at a time when the launch of a new, demanding line of research required support from the fascist leadership. *Nuclei ed elettroni* [Nuclei and Electrons] is Fermi's first popular paper entirely devoted to the problems of the emergent nuclear physics and briefly illustrates the state of the art at the end of 1931. The emphasis is this time on the practical and most visible aspects: the transmutation of elements and the production of energy in nuclear reactions. Fermi writes:

The day one could make the synthesis of nuclei, forming four grams of helium from the reshuffling of the electrons and protons contained in four grams of hydrogen, one would have [...] the loss of about 33 milligrams of mass, which would be transformed into energy, developing about a million kilowatt-hours. [...] It is by no means foreseeable when it will be possible to carry out large-scale nuclear syntheses; and, assuming we succeed to do so in a reasonably short time, it is not possible to say whether the synthesis process will be sufficiently economical to allow a convenient practical application. Until these questions are answered, however, it is not even possible to know that the answer will be unfavorable, and all possibilities should be kept in mind.[38]

The turning point in experimental nuclear physics came in 1934, when Frédéric Joliot and Irène Curie produced some radioactive isotopes by bombarding boron and aluminum nuclei with $\alpha$ particles. The era of artificial radioactivity was beginning: physicists had finally found a way to intervene directly on atomic nuclei and induce nuclear processes. Fermi spoke about this in a 1935 article in the popular science

[36] E. Fermi, La fisica moderna, "Nuova Antologia", 65 (1930), pp. 137–145 (FNM, 62, p. 377).

[37] E. Fermi, Ref. 36 (FNM, 62, pp. 377–378). Under the pressure of war, as we know, the scientists themselves turned into men of practical ingenuity.

[38] E. Fermi, Nuclei ed elettroni, "Gerarchia", 11 (1931), pp. 879–885. This paper is not reported in the bibliography in FNM.

magazine "Sapere",[39] in which—as in his 1923 paper on masses in relativity—he likened nuclear physicists to ancient alchemists[40]:

> The problem of discovering methods to transmute one chemical element into another has been variously studied by several generations of scientific researchers. As is well known, in the Middle Ages, the prevailing part of the work of the alchemists, from which modern chemistry originates, was intensively concerned with attempts to transform mercury into gold; but it is only in very recent times that our knowledge of the structure and properties of the atom has progressed to such an extent that we are able to deal with the problem on a scientific basis and achieve the actual transmutation of a number of chemical elements. And if it is true that many alchemists had the hope of gaining wealth and prosperity, I do not really believe that the same can be said for any of the scientists working in the same field today. The quantity of matter that can be transformed by the most suitable means is so small that it eludes even the most delicate processes of chemical analysis, and only the most sensitive radioactive measurements, which make it possible to detect the presence of even a small number of atoms, have in recent years provided a solid proof of the transmutations that have taken place.

The $\alpha$ particles used by Joliot and Curie—Fermi explained—were not very effective agents for producing the artificial transmutation of elements, because, being positively charged, they could not penetrate heavy nuclei due to the strong electrical repulsion:

> From this one understands the opportunity of looking for the effects of a neutral particle, i.e. the neutron, which is a nuclear-sized corpuscle, with a mass roughly equivalent to that of the hydrogen atom, but without an electrical charge. Although the usable sources of neutrons are far less rich than the sources of alpha particles or protons, one could a priori hope to find compensation for these disadvantages in the fact that neutrons are electrically neutral.

This was the research program that Fermi tenaciously pursued from 1934 onwards, abandoning pure theoretical research—prevalent until then—and becoming a full-time experimentalist. His team—the famous Via Panisperna group (the address of the Physical Institute in Rome)—included Edoardo Amaldi, Franco Rasetti, Emilio Segrè, Bruno Pontecorvo and chemist Oscar D'Agostino. "What I did not know how to do in this experimental technique these friends of mine did know" [7], Fermi said in the seventh Donegani Lecture, in which he recalled for the first time the research he had conducted fifteen years earlier in Rome.[41]

The first result of the systematic work of bombarding the known elements with neutrons was the demonstration of the method's effectiveness: the neutrons did indeed activate both light and heavy nuclei (two-thirds of the sixty or so elements tested).

---

[39] E. Fermi, Trasmutazione artificiale degli elementi, "Sapere", 1 (1935), pp. 55–56. The quotations that follow are taken from this paper.

[40] Alchimia del tempo nostro (Hoepli, Milano) was the title of a popular book on nuclear physics published in 1936 by Ginestra Amaldi (Edoardo's wife) and Laura Fermi.

[41] From the preliminary program of the Donegani lectures (Enrico Fermi Collection, SCRC, University of Chicago Library, Box 53, Folder 9) it can be seen that Fermi had intended to give two technical lectures in Milan on "optical analogies in the properties of neutrons". He then decided to make the first of these two lectures more general, speaking about the research on neutrons carried out in Rome in the 1930s and concluding with an attractive topic, the "age of mummies", i.e. the radiocarbon dating method developed in Chicago by Willard Libby.

"We were all satisfied with the work done—Fermi recalls in [7]—and it was believed that neutrons had given more or less all they could give. But shortly thereafter our group had the good fortune to make one of what are called chance discoveries. It was an experience that frankly I never expected, because being a theoretical physicist I should be able to predict what is going to happen; but instead this accidental discovery happened by chance without anyone going looking for it".

The discovery in question, which took place on 20 October 1934, was that of the effect of slow neutrons.[42] Here is how Fermi reconstructed that event in 1949:

> [...] At that time there was some confusion in the laboratory; one necessarily worked under conditions that were not perfectly reproducible because there were many elements to investigate and sometimes one had to investigate one and sometimes the other in a way that was not always identical in either case. Now, in this somewhat irregular way of proceeding one began to notice differences that seemed inexplicable, and that is, sometimes the activity induced under certain conditions turned out to be small, sometimes it turned out to be larger. It even happened that by placing the plate to be irradiated far enough away from the source, where the activity should have been almost unmeasurable, some activity was measured instead. Since the phenomenon, though strange, could be reproduced, a series of observations began to be made somewhat at random, placing the objects to be activated now here and now there in the hope of catching some circumstance that would give a key to solving the problem; and the key came by chance when it happened that a small piece of paraffin wax was interposed between the source and the object to be bombarded. This small piece of paraffin wax immediately made the intensity increase, albeit by a little, and then the question arose: if a little paraffin wax does this, what will a lot do? A lot was put in, and a lot of paraffin wax actually produced a very large effect. The intensity turned out to be multiplied by a factor of the order of 20 or 30 or 50, so as to indicate that it was indeed a strange phenomenon. Then it was realized that this strange phenomenon was what a theoretical physicist should probably have expected: that is to say it was the phenomenon of neutron slowing [7].

Fermi's explanation of the phenomenon was this: by colliding with the hydrogen nuclei of the paraffin (or water), the neutrons lose most of their kinetic energy and considerably slow down; if one assumes that the activation of a nucleus is the consequence not of a violent collision, but rather of the capture of a neutron, slow neutrons turn out to be much more effective because the probability of their being absorbed by the nuclei is inversely proportional to their speed. Persico notes that this idea "is so far from what a logical type of reasoning could come up with, that only someone like Fermi, with his very acute and perceptive scientific intuition, would dare to suggest it".[43]

By bombarding uranium, the heaviest of the elements then known, the Roman physicists believed they had generated transuranium elements. But in late 1938, two German radiochemistry specialists, Otto Hahn and Fritz Strassmann, discovered the presence of a relatively light element, barium, among the reaction products. This was explained by Lise Meitner and Otto Frisch as the consequence of a process of splitting the initial nucleus into two fragments. Fermi recalls:

---

[42] On the chronology of this discovery, see: A. De Gregorio, Sulla scoperta della proprietà delle sostanze idrogenate di accrescere la radioattività indotta dai neutroni, "Il Nuovo Saggiatore" 19, n. 3–4 (2003), 41–47; F. Guerra and N. Robotti, The Lost Notebook of Enrico Fermi, Springer, Cham 2018.

[43] E. Persico, Ref. 18, Engl. transl., p. 40.

> The splitting phenomenon came to my attention in January 1939. I had just arrived in New York and was somewhat looking around for a field in which to work when news of this discovery came from Germany. Naturally there was a period of excitement in physics: in a dozen laboratories, at the same time, experimenters were rushing to investigate the peculiar properties of this new type of nuclear disintegration. It was then that I realized for the first time that such a phenomenon might one day bring nuclear physics out of the narrow field of pure research and transport it into the field of "big things." I then had the idea confirmed by later experiments, that at the splitting of the nucleus into two fragments neutrons could also be emitted, just as when a drop is broken up one can sometimes see minute droplets spurting out. As a consequence it could be the case, for example, that the first neutron that produced the split produced on average, let us suppose, two: if this were so, of course, there would be the possibility of a chain reaction in which, from one original neutron, an arbitrarily large number could be produced by successive doubling, as long as there remained uranium to be split [7].

The "big things"—the atomic pile, the Manhattan Project and the bomb—kept Fermi busy until 1945. After the wartime interlude, his research resumed on a new front, that of elementary particles. Fermi had already spoken about this in 1933, in a lecture at a SIPS meeting entitled *Le ultime particelle costitutive della materia* [The ultimate constituent particles of matter].

> The fact that in the atomic world we frequently encounter identical objects encourages us to think that the structure of atomic corpuscles is not extremely complex and that, once we have succeeded in analyzing the nature of atomic corpuscles, nuclei and electrons, we do not simply shift the problem of the structure of matter one step further down towards smaller components but, so to speak, reach a sort of landing that, while not representing the ultimate foundation on which the material edifice is built, which is perhaps not attainable by human intelligence, may at least be considered sufficiently solid for a long time.[44]

In those years, the knowledge of the physical world actually seemed to have reached a solid "landing", as Fermi put it. With the discovery of the neutron in 1932, and Heisenberg's and Majorana's theories of nuclear structure, the composition of ordinary matter—electrons, protons, neutrons—was now established. Alongside the electron, Dirac's relativistic quantum mechanics envisaged the existence of the corresponding antiparticle, of equal mass and opposite electric charge: this positive electron (the positron) was discovered in 1933 by Carl Anderson. Quantum field theory also stipulated that fundamental forces were mediated by particles of mass inversely proportional to the range of interaction (the massless photon, for example, was the mediator of the infinite-range electromagnetic force). In 1935, the Japanese physicist Hideki Yukawa proposed that the small-range nuclear force was mediated by a particle with a mass intermediate between that of the electron and the proton, thus called the "mesotron" or "meson". A year later, a particle with this mass was actually found in cosmic rays by Anderson and Neddermeyer. The catalogue of elementary particles seemed definitively established. But it was an illusion, as physicists realized right after the war.

---

[44] E. Fermi, Le ultime particelle costitutive della materia, in "Atti della Società Italiana per il Progresso delle Scienze", 22a riunione (Bari 1933), SIPS, Roma 1934, vol. 2, pp. 7–14 (also in "Scientia", 55 (1934), pp. 21–28) (FNM, 79).

Working under extremely difficult conditions (their experiments began in the basement of a school in Rome during the Nazi occupation),[45] three young Italian researchers, Marcello Conversi, Ettore Pancini and Oreste Piccioni, provided evidence in 1947 that the particle discovered by Anderson and Neddermeyer—the so-called "$\mu$ meson" (or "muon")—interacted weakly with nuclei and therefore could not be the particle predicted by Yukawa[46] [1]. This particle was discovered in the same months in England by César Lattes, Giuseppe Occhialini and Cecil Powell, and was named "$\pi$ meson", or "pion".[47]

The proliferation of particles was just beginning, but it was clear that the "landing" of the 1930s no longer held. Fermi illustrated the situation, with his usual clarity, in the first Donegani Lecture of 1949. "If you were to ask me what is meant by an elementary particle—he observed—I would be embarrassed since the term 'elementary' should be understood in a sense that is rather relative to our knowledge" [1]. And he went on, with an enlightening comparison:

> If one had asked chemists or physicists some fifty years ago whether the atom could be regarded as an elementary particle, probably many of them would have answered in the affirmative, because in the state of knowledge at that time the structure of the atom was not known, indeed it was not even suspected that it could have one. Then, when it was revealed as a complex organism and when this complex nature was more deeply explored, the notion of elementarity was transferred to smaller objects, namely the nucleus; but today even this has revealed much of its complex nature. In general, therefore, it could be said that at every stage of science, particles whose structure is unknown are called elementary, and therefore can be considered as points [1].

And about the increasing number of particles observed:

> If a physicist or chemist who knew nothing about atomic structure began to explore the nature of chemical objects and discovered a first atom, for example, an iron atom, he might suppose it to be an elementary particle. But then if, after this atom of iron, the chemist discovers an atom of another nature, for example of sulfur, and then an atom of oxygen and so on (of course the order I give you is absolutely arbitrary), logically the confidence of this researcher in the elementarity of the atom would gradually diminish because such a large number of particles would contradict the very concept of elementarity. Even for the elementary particles known today, a situation not entirely dissimilar to the one just now mentioned is taking place [1].

The particles that Fermi considered as "well established" in 1949 were the electron, the positron, the proton, the neutron, the pion (in the two versions, with positive and negative electric charge), the muon (also positive and negative), and the photon. To these he added, "with some reservations", six other particles, whose existence "is suggested either by theoretical reasons that are not absolutely conclusive, or by

---

45 See E. Amaldi, Da via Panisperna all'America, G. Battimelli and M. De Maria eds., Editori Riuniti, Roma 1997.

46 The implications of these experimental findings were discussed in E. Fermi, E. Teller e V. Weisskopf, The Decay of Negative Mesotrons in Matter, "Physical Review", 71 (1947), pp. 314–315 (FNM, 232).

47 For the early history of particle physics see L. M. Brown and L. Hoddeson (eds.), The Birth of Particle Physics, Cambridge University Press, Cambridge (UK) 1983.

experiments that are far from complete and reliable at present time". These hypothetical particles were the neutrino, the graviton (the mediator of the gravitational force) and four others not explicitly mentioned: from the preparatory notes to the lecture,[48] we know that they were the antiproton, the antineutron, the "neutretto" (i.e. the neutral pion) and the $\tau$ (corresponding, in today's nomenclature, to the K meson).[49] The neutrino had not yet been observed, but Fermi assumed its existence, "at least as a working hypothesis" [2]. In one of the Silliman Lectures the following year he joked about this: "Since the nature of the neutrino is such that its existence cannot be known, no one can prove that physicists were wrong in inventing it".[50] As for the antiproton and antineutron, he was convinced of their existence by virtue of a general symmetry principle of quantum field theory.[51]

Even assuming that some of these particles were not real, their number appeared to Fermi sufficiently large "to cast doubt on the elementarity of some of them". "At this point—he said—the need arises for a systematics to be carried out, whereby we can understand why some of these particles are observed in nature, and others are not. This is, at the present stage, a very logical but entirely unfulfilled desire, since no one has so far succeeded successfully in this intent".[52]

Fermi did not just talk about the new particles; he also showed them. As early as the 1920s, he had emphasized the usefulness of Wilson's cloud chamber, which, in addition to being a fundamental research tool, had made it possible to see the elementary constituents of matter with one's own eyes.[53] By the 1940s, another powerful technique for the direct observation of particles had been developed, based on photographic emulsions. To his audience, Fermi showed a dozen pictures of particles and subatomic processes, from the positron to the decay of the pion. "I wanted to show some photographs [...]—he said—because perhaps some of you might have thought that everything discussed resulted only from theory. Instead, as it turns out, these are facts experimentally proved by ordinary materials and means, such as the photographic plate" [1].

At the end of his stay in Rome, Fermi visited the Italian National Research Council (Consiglio Nazionale delle Ricerche), where he praised Italian research in particle physics, based essentially on the study of cosmic rays. On the other side of the Atlantic, the race to build larger and larger accelerators had already begun. The

---

[48] Enrico Fermi Collection, SCRC, University of Chicago Library, Box 53, Folder 9.

[49] In the last Donegani lecture [9] Fermi spoke about the magnetic monopole, "a particle that probably does not exist ...but if it did exist it would be beautiful", as he said in [7].

[50] See Fermi expounds upon speculative particle theory, "Yale Daily News", April 13, 1950, p. 3.

[51] See E. Fermi, Elementary Particles, Ref. 17, §23.

[52] We know from the preliminary program of the two lectures on elementary particles (Ref. 51) that Fermi intended—if there had been time—to discuss the possibility of the pion being a composite nucleon-antinucleon state, according to the model he had proposed two months earlier in E. Fermi and C.N. Yang, Are Mesons Elementary Particles?, "Physical Review", 76 (1949), pp. 1739–1743 (FNM, 239). But this topic was not dealt with.

[53] E. Fermi, Problemi attuali della fisica, "Annali dell'Istruzione Media", 5 (1929), pp. 424–428. See also E. Persico, Una finestra sul mondo atomico: la camera di Wilson, "Atti della Società Italiana per il Progresso delle Scienze", 23a Riunione (Napoli 1934), SIPS, Roma 1935, vol. 2, pp. 89–100.

Americans, said Fermi, are "a people who, by nature and tradition, love machines, and this love, in the field of science, manifests itself in constructions that have something of the Egyptian pyramids. [...] There are several new giant cyclotrons that weigh around 2000 tonnes. And other machines whose names end in '...tron'. The most spectacular one now under construction is the bevatron, which means machine to produce billions of electron volts".[54]

In the final years of his life, Fermi had at his disposal one of these accelerators, the 450 MeV Chicago synchrocyclotron, with which he performed a rich research campaign on pion-nucleon scattering. He spoke about this on his last visit to Italy, in the summer of 1954, in a series of memorable lectures in Varenna, on Lake Como. His collaborator Bernard Feld recalls him as follows:

> Here was Fermi at the height of his powers, bringing order and simplicity out of confusion, finding connections between seemingly unrelated phenomena; wit and wisdom emerging from lips white, as usual, from contact with chalk, in that clear, resonant voice that had never lost the soft Italian vowel endings on a perfectly colloquial American delivery.[55]

It was the last time Enrico Fermi enchanted his audience.

---

[54] La visita di Enrico Fermi al Consiglio Nazionale delle Ricerche, "La Ricerca Scientifica", 19 (1949), pp. 1113–1118. It was at the Bevatron that Emilio Segrè discovered the antiproton in 1955, and Bruce Cork and Oreste Piccioni discovered the antineutron in 1956.

[55] B. Feld, in FNM, vol. II, p. 1004.

# Presentation of Enrico Fermi, Lectures of Atomic Physics edited by Professors and Assistants of the Universities of Rome and Milan in 1949

**Guido Castelnuovo**

*In October 1949 Enrico Fermi, accepting an invitation extended to him by the "Donegani Foundation," gave six lectures in Rome and three in Milan on various topics in atomic physics, lectures that were followed with lively interest by a large audience that crowded the lecture hall of the Physics Institute in Rome and the large hall of the Montecatini Society in Milan. These lectures, recorded and shorthanded, were then edited by a few professors and assistant professors in physics at the Universities of Rome and Milan; each lecture bears the name of the editor. In presenting today the volume containing these lectures to the public, the Accademia Nazionale dei Lincei, to which the "Donegani Foundation" is attached, wishes to renew the expression of its gratitude to the distinguished lecturer who, after eleven years of absence, has made his voice resound among us and has demonstrated his interest in the young Italian school of physics, to which he had given the first impulse. The Accademia thanks the Montecatini Society, which with its generous contribution made it possible to implement at once the noble intentions in favor of science expressed by Guido Donegani in his will. Finally, the Accademia thanks the professors and assistant professors who, with care and love, undertook the no small task of adapting for print lectures initially prepared for oral exposition. Our Academy trusts that this volume will contribute to the progress of modern physics in our country and abroad.*

---

Guido Castelnuovo, President of the Accademia Nazionale dei Lincei in 1949.

G. Castelnuovo
Accademia Nazionale dei Lincei, Roma, Italy



A. Treves and P. Tucci (eds.), *Enrico Fermi, Atomic Physics Lectures*, History of Physics,
https://doi.org/10.1007/978-3-031-68690-0_3

# First Lecture The Elementary Particles (Part One)

**Enrico Fermi**

## Edited by S. Sciuti

In accordance with the program that has been agreed with the Donegani Foundation, of the six lectures I will give here in Rome, the first one will be of a rather general nature, while the other five will be of a rather specific nature, devoted particularly to physicists.

The topic I will discuss today concerns "elementary particles". But if you were to ask me what is meant by an elementary particle, I would be embarrassed since the term "elementary" should be understood in a sense that is rather relative to our knowledge.

If one had asked chemists or physicists some fifty years ago whether the atom could be regarded as an elementary particle, probably many of them would have answered in the affirmative, because in the state of knowledge at that time the structure of the atom was not known, indeed it was not even suspected that it could have one. Then, when it was revealed as a complex organism and when this complex nature was more deeply explored, the notion of elementarity was transferred to smaller objects, namely the nucleus; but today even this has revealed much of its complex nature. In general, therefore, it could be said that at every stage of science, particles whose structure is unknown are called elementary, and therefore can be considered as points.

Within the framework of these very general considerations, another one can also be made, which concerns the number of particles. If a physicist or chemist who knew nothing about atomic structure began to explore the nature of chemical objects and

(*) Held on October 3rd, 1949 in the lecture hall of the Physics Institute of University of Rome.

E. Fermi (✉)
University of Chicago, Chicago, IL, USA

 
A. Treves and P. Tucci (eds.), *Enrico Fermi, Atomic Physics Lectures*, History of Physics,
https://doi.org/10.1007/978-3-031-68690-0_4

discovered a first atom, for example, an iron atom, he might suppose it to be an elementary particle. But then if, after this atom of iron, the chemist discovers an atom of another nature, for example of sulfur, and then an atom of oxygen and so on (of course the order I give you is absolutely arbitrary), logically the confidence of this researcher in the elementarity of the atom would gradually diminish because such a large number of particles would contradict the very concept of elementarity. Even for the elementary particles known today, a situation not entirely dissimilar to the one just now mentioned is taking place.

The first elementary particle is, as you know, the electron, which was discovered about fifty years ago and is now so familiar that its name is commonly found in the daily press. Later the proton was discovered, and so the elementary particles were for some time two; then again the neutron was discovered, thus arriving at three elementary particles.

The elementary particles that we can, at the present time, consider as well established are ultimately as follows:

- 1st the electron (normally negative, with its positive electron variety);
- 2nd the proton;
- 3rd the neutron;
- 4th the $\pi$ meson (divided into two varieties, one with positive electricity and one with negative electricity);
- 5th the $\mu$ meson (also in the positive and negative varieties);
- 6th the photon.

These, with all their varieties, are nine particles and constitute a large enough number to cast doubt on the elementarity of some of them.

At the present time, however, there are not many indications, either experimental or theoretical, to decisively discard this hypothesis, and so we must settle, at least tentatively, for the above pattern.

To finally complete the list of elementary particles, another half dozen should be mentioned, which I present, however, with some reservations because they cannot be considered well established. Each of them has only a 50% probability and therefore only three may perhaps survive future research.

Among these particles there is the neutrino, which has so far always escaped all experimental investigation and whose existence has been established theoretically to account for the disappearance of a certain amount of energy in some processes. This energy shortfall would have invalidated the principle of conservation of energy and momentum (against the traditional belief that their validity is unquestionable) had it not been precisely postulated that the amount of energy under consideration did not disappear but was instead transferred to an electrically neutral particle, probably massless and extremely difficult to observe experimentally. Today, however, there are reports of experiments that will perhaps lead to its discovery, assuming it actually exists.

Then there is a very second-order particle, called the graviton, which would be the "quantum" of the gravitational field. This is a particle that most likely exists, assuming that our theoretical ideas about how these phenomena should be interpreted are correct. In any case, it will be very difficult to observe it directly, because it is almost as unobservable as the neutrino.

I will not name the other four particles in detail; I will only mention that their existence is suggested either by theoretical reasons that are not absolutely conclusive, or by experiments that are far from complete and reliable at the present time.

All in all, we can say that, considering three of the names in the second list to be fairly probable particles, we have altogether about a dozen elementary particles. This is too large a number to consider the present stage to be definitive.

At this point the need arises for a systematics to be carried out, whereby we can understand why some of these particles are observed in nature, and others are not. This is, at the present stage, a very logical but entirely unfulfilled desire, since no one has so far succeeded successfully in this intent. However, I would not like to discuss at this time the theoretical problems connected with these particles, but rather to talk about some of the experimental methods and circumstances under which the particles are observed and produced.

The methods of observation have been largely known for a long time. Therefore, I will not describe them in particular detail. Everyone has heard of the Geiger and Müller counter as a method of detecting ionizing particles; of the Wilson chamber, as a method in which these particles leave along their path a trail that can be photographed.

In more recent years new methods have been developed: one of them is the photographic method, which is based on an extremely simple principle, but which has required, in order to reach today's development, a very considerable refinement of technique. An ionizing particle passing through the gelatin of a photographic plate produces by ionization a small blackening track. This, in order to be observed, requires a magnification of 1000–2000 times.

From the examination of these tracks, namely, from their geometric behavior and density, conclusions can be drawn as to the nature of the particles.

Another method that has been gaining ground in recent years is one that, in principle, was already being used in the early days of radioactivity studies, particularly by Lord Rutherford and the Curies. It is the method of so-called scintillations, based on the fact that a particle, by etching over a special screen, produces a scintillation visible to the microscope. Formerly, scholars spent hours and hours at the microscope, in the laborious endeavor of observing and counting these scintillations. Today, however, an electron-multiplying photoelectric counter is substituted for the eye, so that the uncertain and painful part of yesterday is now done automatically, with great advantage in time and accuracy.

By way of introduction, I would like to show some slides now, to show what these elementary particles look like to the observer.

Figure 1 represents a Wilson chamber photograph of a fast electron: the track produced by this particle consists of a series of droplets which give a clear indication of the passage of a high-energy electron passing through the chamber field, without undergoing any deviation. Also noticeable are some very pronounced tracks due to slow electrons; these collide with the gas molecules through which they propagate and undergo a series of deflections, giving the track a twisted appearance. It can also be seen in this photograph, that the fast electron track is much thinner and therefore contains a smaller number of droplets than that related to the slow electron track. This is simply due to the fact that the latter produce a more intense ionization than the others.

**Fig. 1**

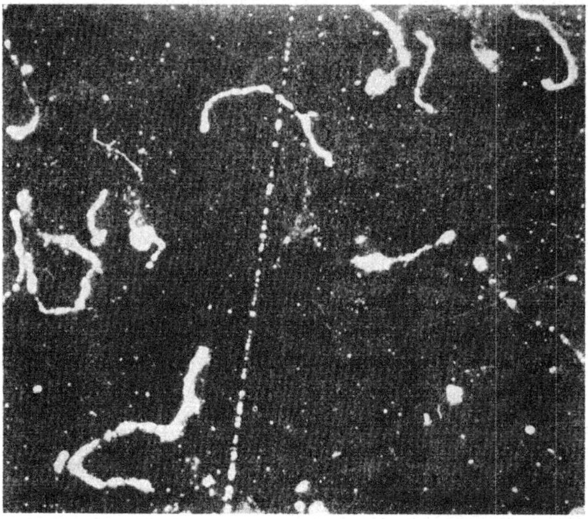

In Fig. 2 one can see the track of a particle (it is not an electron but a particle that is similarly observed in cosmic rays) colliding almost centrally with an electron, which is jolted forward along an almost circular trajectory. This is because in this, as in most experiments of this kind, the Wilson chamber is immersed in a strong magnetic field for the very purpose of producing deflections. From a quantitative study of them one can, in many cases, calculate the velocity of the particles under examination.

**Fig.2**

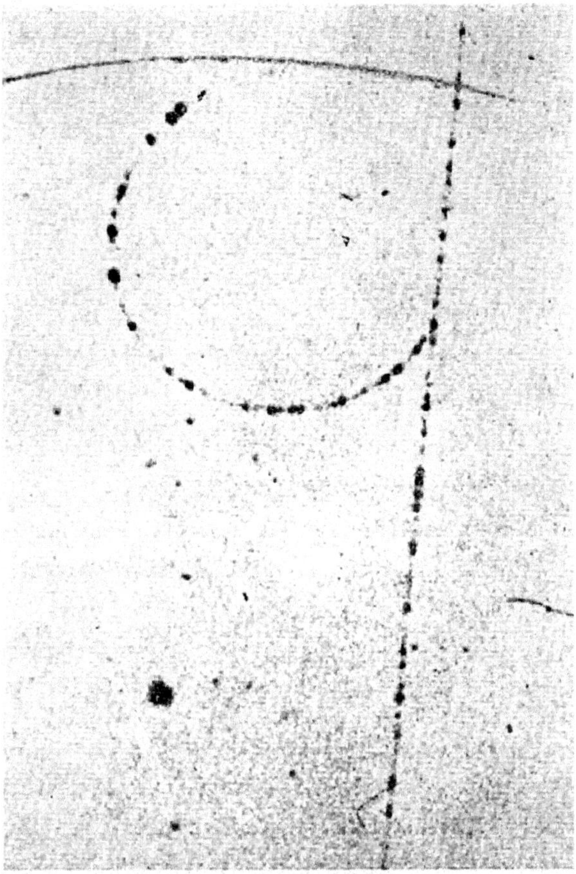

**Fig.3**

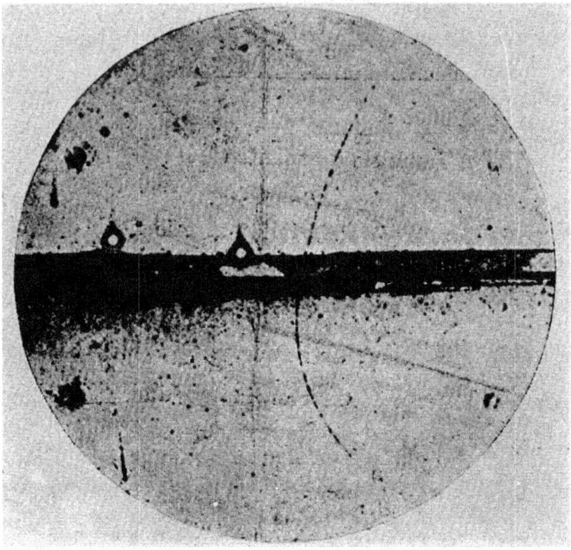

Fig.4

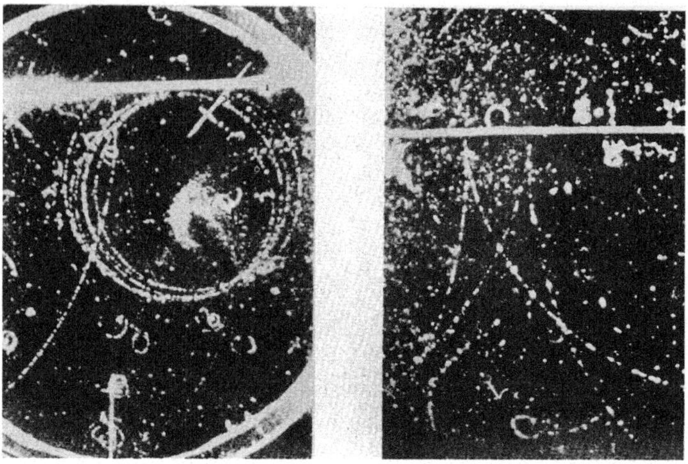

Figure 3 shows a positive electron entering the chamber and traversing a sheet of material and losing energy, as can be seen from the fact that its trajectory, a little curved in the first part, is appreciably more so after passing the absorber.This means that the particle has lost velocity, because in traversing the screen it has lost energy.

In Fig. 4, the phenomenon of pair formation (positive and negative electrons) by gamma rays is presented three times. Some $\gamma$ rays, not visible in this type of observation, arrive on a sheet of material and create an electron pair whose trajectory is clearly visible. In particular, it can be seen that the trajectories of each pair are curved in opposite directions, which corresponds precisely to the fact that the two electrons in each pair have opposite charges. In this figure, one can also see a track that is wrapped in many circles. It is due to a positive electron that is so slow that it is curved by the applied magnetic field, according to a circle of fairly small radius.

Figure 5 shows more complicated phenomena, which I am not going to describe in detail and which are observed in the study of cosmic radiation. These are the so-called showers, and a very complex phenomenon, which is not entirely understood even today, and which consists partly of a formation of positive and negative electrons.

It is interesting to note from a technical point of view that this figure consists of two essentially similar photographs: they were taken with two cameras placed at right angles to each other and can, with appropriate methods, be treated as stereoscopic photographs, in order to reconstruct the actual nature of the phenomenon that occurs, not only in the plane of projection but also in space.

Figure 6 shows another shower (note that this photograph was taken without a magnetic field so the tracks appear straight), which I submit for your attention simply to demonstrate how these phenomena, when revealed by appropriate means, are truly spectacular.

I will now show two other photographs in which a very remarkable phenomenon is presented and that is the phenomenon of the transformation of one of the elementary particles into another—the conversion into an electron of the so-called $\mu$ meson. At a certain point it stops in the gas of the chamber, but it does not stay there for a long time; in fact after one or two millionths of a second it will spontaneously transform into another particle and that is to say into an electron.

**Fig.5**

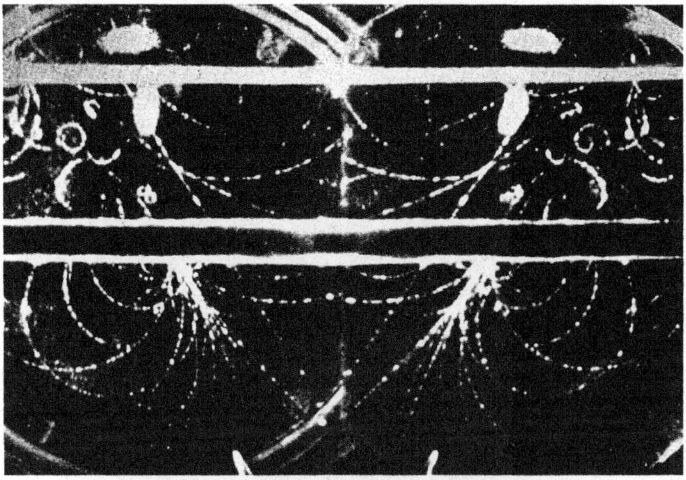

**Fig.6**

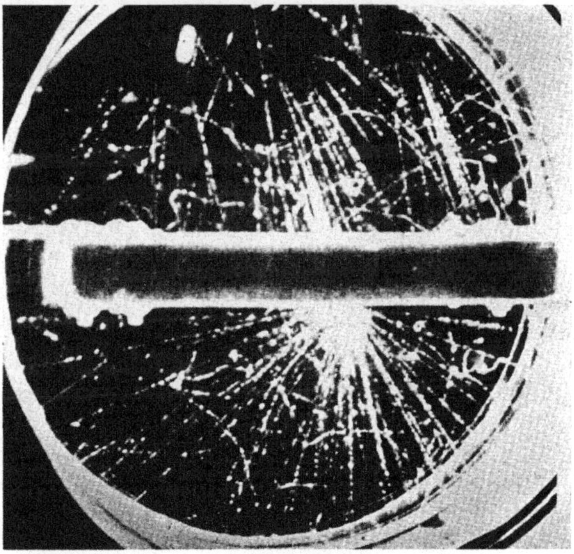

Fig.7

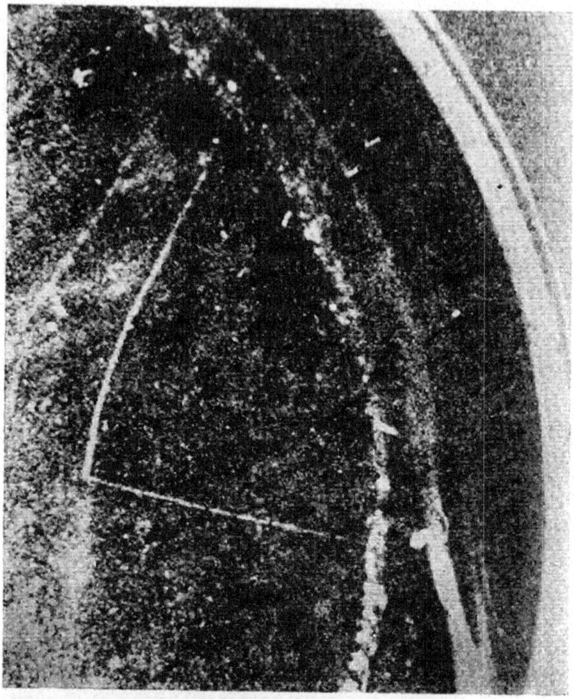

Fig.8

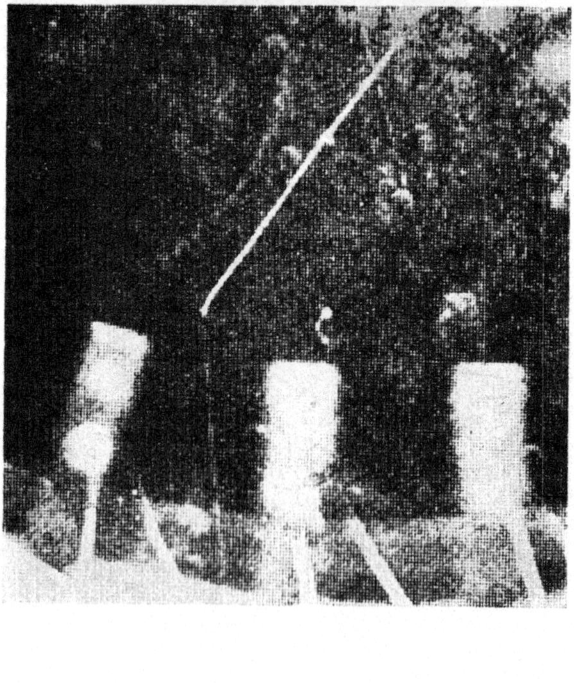

In fact, you see in Figs. 7 and 8 the track of a $\mu$ meson, which can be recognized by various characteristics, particularly by the high ionization density and also by some uncertainty of the track (which indicates that the particle undergoes fairly frequent collisions with gas molecules). At a certain point, the particle stops and after a few millionths of a second dies, transforming into an electron which gives a thinner track.

Where are these elementary particles produced?

Of course, I would not like to talk about electrons, which are so well known to most that they have lost all novelty. Rather, I would like to talk about newer particles and particularly mesons.

The most common laboratory in which these particles are manufactured is cosmic rays. When these arrive in the upper atmosphere they produce, by colliding with the nuclei of the air or the nuclei of the various materials with which observations are made, disintegrations in which in fact particles such as mesons are produced.

To make this more obvious, I would like to show some more photographs (Figs. 9 and 10) of such phenomena. They were obtained by strong magnification to the order of ten to one hundred thousand times and reproduce phenomena occurring within the gelatin of a specially prepared photographic plate.

These disintegrations have the appearance of stars with a large number of rays, each representing the track of one of the disintegration products of a nucleus.

In Fig. 9, the incoming projectile, coming from a vertical direction, impacted a silver nucleus at some point, producing a large explosion in which nucleus fragments are projected in all directions. Among these beams some have the characteristics of heavy particles (very marked traces) others correspond to particles of high velocity and small charge; probably several mesons are among the latter.

Fig.9

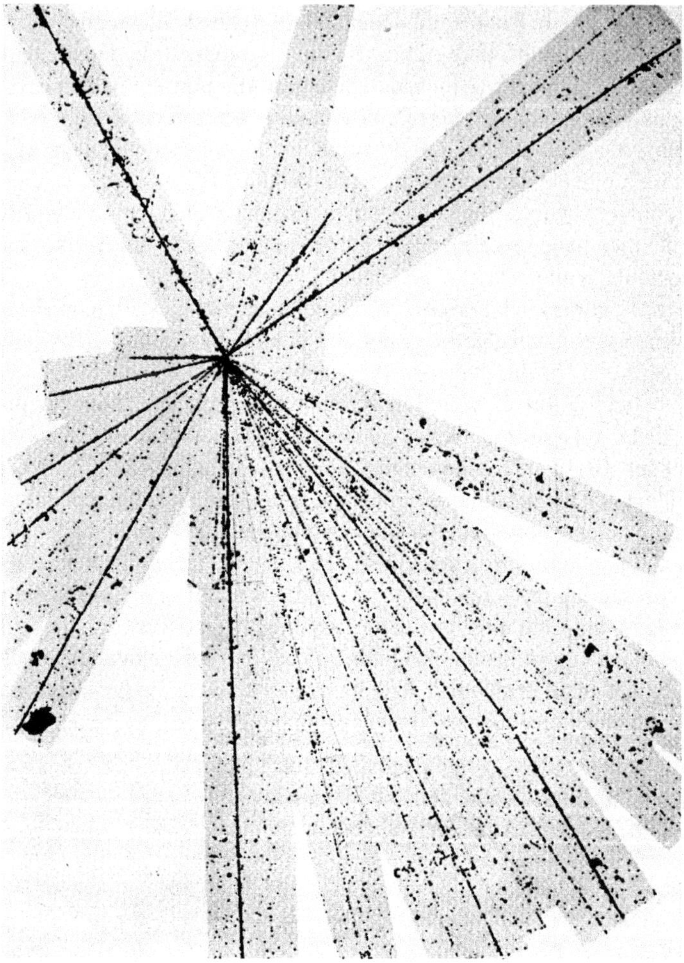

In Fig. 10 we see a star produced by a fast particle breaking a nucleus; we note in particular a ray which has such a long path that, to make it fit into the dimensions of the photograph, it had to be divided into two parts, the second of which is reproduced in isolation below the first. It is produced by a beryllium nucleus, which jumps out from a heavier nucleus.

I wanted to show some photographs (of course, hundreds could be shown, collected in the various cosmic ray laboratories and in particular in the Testa Grigia laboratory) because perhaps some of you might have thought that everything discussed resulted only from theory. Instead, as it turns out, these are facts experimentally proved by ordinary materials and means, such as the photographic plate.

**Fig.10**

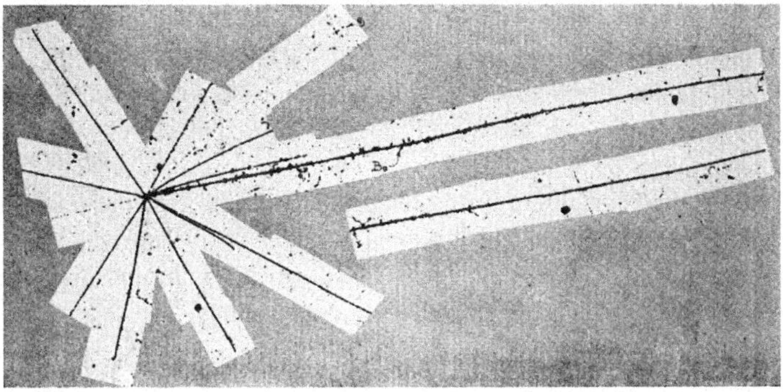

I would now like to explain a little more specifically the research that led to the discovery of the meson. By the way, this particle is called a meson because by this it is meant to indicate that its mass lies between that of the electron, which is lighter, and that of the proton, which is heavier. The mass of the meson corresponds to about three hundred times that of the electron.

The idea of the meson was first introduced in 1935 from theoretical consideration by a Japanese physicist, Yukawa, who "invented" this particle, or rather proposed a theory in which he hypothetically admitted its existence to explain the nature of nuclear forces. Time does not permit me to describe the theory in detail; but rather I would like to put forward some points of view that are of interest. In particular, I would like to justify why a particle, with a fairly considerable mass such as the meson, is well suited to explain small-range forces such as nuclear forces.

Let us first refer, as a basis, to the electromagnetic field. Precisely, if we consider two particles A and B, at a certain distance from each other, the electric forces to which they are subjected are transmitted by a field, and it can be said that particle A produces around itself an electric field within which particle B will come to be, which in this way is subjected to its action (and vice versa). The field is thus the intermediary that transmits the force between the two particles.

Now electrical forces have a peculiarity, that is, they are forces of long range, reaching a great distance; they do indeed decrease with distance, but only with the square of it, that is, with a fairly slow law, so that even distant particles interact with each other. But not so the nuclear forces, which are of such a nature that they can act only if the two particles are very close together (and, when I say very close, I mean at a distance of the order of magnitude of $10^{-13}$ cm, that is to say, at a distance a hundred thousand times smaller than the radius of the atom, which is of the order of $10^{-8}$ cm); in a word, nuclear forces are short-range forces. The mechanism, however, can be regarded as analogous to that carried out for the electromagnetic field, in other words one can imagine that even such short-range forces are transmitted by a field, which will not, of course, be an electric field, but a different kind of field and the nature of which we now have to clarify. Just as the electric field has its photons, which are, as it were, quanta of electromagnetic energy and which behave in all respects like particles, so, this being a general property of all fields, the field of nuclear forces will have its analog of photons, which we shall call mesons.

What will the characteristics of these quanta be so that the forces come to have a small range? Yukawa's answer is that they must have a relatively large mass. The reason for this can be understood almost without formulas, at least for those familiar with the methods of quantum mechanics. Consider two particles, a proton and a neutron, at a certain distance from each other; each of them will surround itself with its own field which means emitting quanta of this field and thus the particles will emit mesons; but to emit a meson costs energy and if it has a mass M, the energy required for its creation alone will be given by:

$$\mathrm{E} = \mathrm{M}c^2 \qquad [c \text{ speed of light}] \tag{1}$$

In quantum mechanics, one can, in a sense, borrow energy, provided, however, that the term is very short. Practically, if one wants to borrow a certain amount of energy, say for example $\mathrm{M}c^2$, according to Einstein's relation, this can be done but only for a time of the order of magnitude given by:

$$t = \frac{h}{\mathrm{M}c^2} \qquad [h \text{ Planck's constant}] \tag{2}$$

From (2) it can be seen that if a large energy is to be taken on credit, the credit term must be short because the energy being taken on credit is in the denominator; (2) then represents the time the meson can stay outside the nucleus in free space. If we then assume that the speed of this free meson is the maximum speed at which a particle can move, i.e., the speed of light, we see that the maximum distance to which it can go before it is called back to pay off the debt is given, as an order of magnitude, by the product of time times the maximum speed at which the particle can move:

$$d = \frac{\mathrm{h}}{\mathrm{M}c}. \tag{3}$$

Thus, (3) expresses, apart from some numerical coefficients, the range of action of nuclear forces. If nuclear forces are then to be represented, it must be admitted that the distance (3) must be precisely of the order of magnitude of $10^{-13}$ cm (which is the experimental value of the radius of action of these forces), and since both Planck's constant and the speed of light are known, the mass of the meson or at least its order of magnitude can be derived. Yukawa, who devised this theory and developed it mathematically in a very elaborate and comprehensive way, came to the conclusion that if there was anything true, at least qualitatively (he was not sure because both at that time and today we cannot give a clear interpretation to these phenomena) there had to be certain particles, whose mass, calculated essentially on the basis of the reasoning now made, had to be of the order of magnitude of 200 or 300 times the mass of the electron.

Shortly thereafter two groups of physicists studying cosmic rays (precisely Anderson's group in California and Stevenson's group in Massachusetts) found particles in their plates that had a mass about 200 times the mass of the electron. They concluded that these were the Yukawa particles. However, later it could be clarified that this was not so because it was ascertained that the newly discovered particles had nothing to do with them. The first clear indication that the new particles were not Yukawa's mesons came from an experiment that was conducted here in Rome by Conversi, Pancini and Piccioni (the work was done, at least in good part, in a cellar during the period of the resistance, while these guys were in hiding for political reasons). The result of this experience can be roughly summarized as follows. A positive or negative meson stops in a block of matter. If positive, it will not be able to enter nuclei especially when stationary because electrical repulsion keeps it outside of them, leaving it practically in free space. Here, after a time of the order of a millionth of a second it disintegrates by emitting an electron as in Figs. 7 and 8. If the meson is negative, it will be attracted to the nuclei of the atoms in the midst of which it stops, and in that case it will fall very quickly into one of them. If this meson has the same characteristics as the Yukawa particle, and that is to say, if it is precisely the quantum that determines nuclear forces, the interaction with the nuclei will be so violent that it will be absorbed by them in a time of the order of $10^{-22}$ s.

Well, the experiments of Conversi, Pancini, and Piccioni showed that the meson does indeed disappear under particular circumstances, but this is after a relatively large amount of time, thus revealing a rather small interaction with nuclei: this is in fact about a million million times smaller than that expected from the Yukawa meson. This is therefore a difference between the theoretical predictions and the experimental result, which evidently cannot depend on errors in detail.

Yukawa's entire theory was then questioned not only in matters of detail but also in its general basis, and this state of affairs persisted until the solution to the riddle came from the work of Powell and Occhialini (Bristol) in which they pointed out the fact that there are two species of meson, one with the characteristics of Yukawa's meson, and another that is more commonly observed in cosmic radiation: I say more commonly, because both types are now observed. These authors also show that the first type i.e. the $\pi$ meson spontaneously converts into the other meson ($\mu$ meson), as I will show with two photographs.

Figure 11 depicts a star that, at first glance, appears very similar to the previous ones; in fact it offers an interesting detail, namely, a track (marked by the letter $\pi$) that from examination of all features appears to belong to a meson. At the point where this particle stopped, nuclear disintegration occurs, revealing a strong interaction of the meson with the nuclei: it is thus a particle corresponding to Yukawa's theory.

In Fig. 12 we see the track of another $\pi$ meson, which, having a positive charge, cannot interact with nuclei because of Coulomb repulsion. At a certain point it stops and disintegrates giving rise to a meson of the other type, that is to say, a $\mu$ meson which, after traveling a certain distance in the gelatin (a rather long trace) stops and disintegrates in turn into an electron.

Fig.11

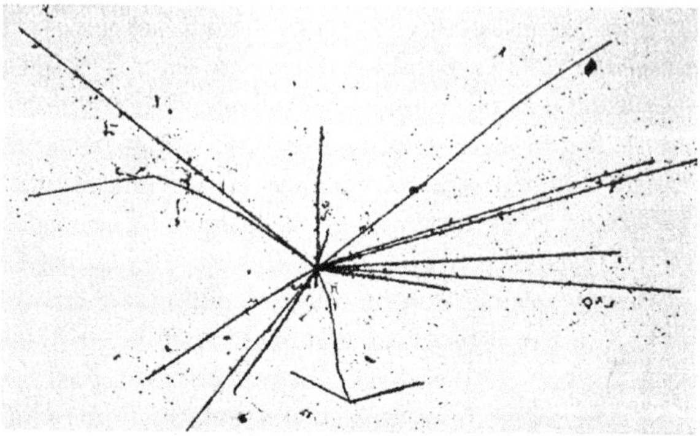

Fig.12

From what I have said, it appears that any hope of understanding in some detail what are the properties of the atomic nucleus (which at present we know purely empirically but are unable to calculate or understand in detail) is distinctly tied to the study of the properties of the $\pi$ meson. Probably its properties will be quite similar to those indicated in Yukawa's theory, or more properly in "Yukawa's theories" since there can be several (at the present time there is no way to decide which of them is the good one, assuming one is, because we do not yet know enough about the nature of these phenomena).

I would like to add a few more words on the question of elementary particles, which, as I said, are particles without structure or at least with a structure that we do not know. But in fact many of them reveal some structure, so that any right to consider them elementary comes to an end. For example, in the scheme previously described, in which a proton surrounds itself with its field and emits mesons, the proton particle should not actually be considered, but rather a point with two particles occasionally arriving at a very small distance from each other but not infinitesimal. I will also say the same for more common and familiar particles, such as the electron. The latter is surrounded by an electromagnetic field which essentially represents a structure and which, more or less modified, accompanies it in all its movements. This state of affairs, i.e., the physical electron, could be described as a mixture of a certain idealized, point-like electron and a cloud of photons surrounding it that represent the electromagnetic field. So these what we still continue to call elementary particles are actually complicated objects. In trying to analyze this complication we find considerable difficulties for the theory. For example, we find that the electromagnetic mass, which is the mass due to the field surrounding the electron, turns out to be infinite.

And so we arrive painfully at what is now the frontier of our knowledge in the field of elementary particles. It is a naturally moving frontier, a fluid frontier, a frontier that is likely to be consolidated as new experimental data are obtained in laboratories, data that will serve to direct the thinking of theoretical physicists toward a more circumscribed field.

And when this happens, a great advance in physics may be made, because it is presumable that the resolution of this problem will be accompanied by the resolution of the problem of nuclear forces and will therefore make it possible to make a true theory of the nucleus and not an empiricism, as we have today, in which we know how to use the nucleus but do not know how it is made.

But of course beyond this frontier there will be more and more, because it seems to be the fate of science that each group of discoveries, expanding the area of knowledge, must also expand the area of things unknown!

## Biographical Notices on Sebastiano Sciuti

### AT and PT

Sebastiano Sciuti was born in Naples in 1917. In 1935, fascinated by a lecture by Orso Mario Corbino (1876–1937) on the first experiments in neutron physics, he enrolled in Physics at Naples university. There, when he was already preparing his thesis on experimental physics, he followed the lessons of Ettore Majorana (1906–1938), who had just settled in Naples. He graduated in 1938 and started working under Antonio Carrelli, at the time dominus of physics in Naples.

During his long military service from 1939 until 1945, as a graduate in physics, he was assigned to the Istituto Superiore per le comunicazioni (Higher Institute for Communications) where he met Oreste Piccioni (1915–2002) and became a good friend of his. Piccioni introduced him to the Roman physics environment where he met Gilberto Bernardini (1906–1995) and Edoardo Amaldi (1908–1989) and was introduced to the cosmic ray group. In 1945, at the end of the war, he reestablished his contacts with the group.

In fact in his autobiographical Notes, he remembers collecting, with Pancini, the Fermi conferences delivered at the University of Rome, which are presented in this book. As he recalls in one of his articles he had no relationship with Fermi except when he edited one of the conferences (the first) misplacing the order of the slides by putting them in all possible combinations and the last attempt being the good one!

From 1956 he was involved in the Comitato Nazionale per le Ricerche Nucleari (CNRN) (National Committee for Nuclear Research) and contributed significantly to the birth of Applied Nuclear Physics in Italy. The CNRN was subsequently transformed into the Comitato Nazionale per l'Energia Nucleare (CNEN) (National Committee for Nuclear Energy) first and then Energia Nucleare ed Energie Alternative (ENEA) (Alternative Energy and Nuclear Energy).

At the end of the sixties he shifted his interests to new physical technologies for non disruptive analysis of pieces of art, and in particular for in situ study of celebrated paintings. His studies of Michelangelo's frescoes in Cappella Sistina are very well known.

He died in 2016.

For further information see: Bonolis (2008); Sciuti (2001); Spezia (2016).

## References

Bonolis, Luisa (2008) Maestri e allievi nella fisica italiana del Novecento (Pavia: La Goliardica Pavese 2008).

Sciuti, Sebastiano (2001). "Enrico Fermi divulgatore e persuasore scientifico" Il Nuovo Saggiatore 2001 5-6: 23-25.

Spezia, Ugo (2016). "Ricordo di Sebastiano Sciuti" Scienza e Tecnica 2016 537: 9-10.

# Second Lecture The Elementary Particles (Part Two)

**Enrico Fermi**

## Edited by L. Mezzetti

In the previous lesson we listed a number of elementary particles; today we will talk a little more in detail about their properties, or at least some of their properties.

These are many. Some of them are known, some of them are unknown; perhaps, indeed certainly, most of them are unknown. But there are certain facts which are either known, or suspected, and which make it possible, through a mathematical analysis, to begin the construction of a theory of them. I say begin, and I want to emphasize this expression, because we are extremely far from finishing it.

Perhaps the most remarkable peculiarity that results from the experimental study of these particles lies in the fact that, in various cases, they transform into each other; or, in other cases, particles that were previously unobservable suddenly appear. This is not an entirely new fact in the physics of quantum phenomena; and perhaps the first case in which it presented itself clearly is that of the emission of light. This phenomenon was interpreted in classical physics as a gradual emission of waves due, shall we say, to the shaking of the electromagnetic field by the electrons of the atom. But, from the quantum point of view, the emission of light takes place instead by "quanta". At a certain moment an atom emits a quantum but it cannot be said in any way (and the analysis that is done mathematically of this phenomenon does not indicate in any way) that the quantum was pre-existing in the atom and at a certain point comes out of it. Instead, it must be thought that the quantum is formed at the

(*) Held on October 5th, 1949 in the lecture hall of the Physics Institute of University of Rome.

E. Fermi (✉)
University of Chicago, Chicago, IL, USA



A. Treves and P. Tucci (eds.), *Enrico Fermi, Atomic Physics Lectures*, History of Physics,
https://doi.org/10.1007/978-3-031-68690-0_5

same moment when the quantum transition between the two energy levels of the atom takes place and light is emitted.

This phenomenon of light emission can therefore be indicated schematically by an equation, which is the model for other equations of the same type, which will be written in a moment:

$$e \rightarrow e' + \gamma$$

where we indicated with the letter $e$ the existence of an electron in a certain orbit (energy level) that at a certain time transforms into an electron $e'$ located in a different orbit and a quantum of light, which we indicated with the letter $\gamma$. Where before there was only one elementary particle (an electron), another one has appeared: a quantum of light.

As new elementary particles have been discovered and as our knowledge of them, though incomplete, has gradually deepened, new reactions, similar to the one now written, have been added, so that we now know about ten or so in all. In some cases, these reactions are known incompletely, because some particles are seen to come out, but others are suspected to come out that are not observed. In general, the particle that is suspected to come out, even though it is not observed, is the neutrino.

In the remainder of this discussion we will certainly assume that the neutrino exists, making an assumption that is actually not particularly justified so far by experimental facts but which can probably be used, at least as a working hypothesis.

As an example of an incompletely known reaction, let us take the reaction that takes place in $\beta$-ray emission. $\beta$ ray emission consists essentially of the transformation of a neutron N from the nucleus into a proton P, with the emission of an electron $e^-$ and a neutrino $\nu$, according to the following scheme:

$$\mathrm{N} \rightarrow \mathrm{P} + e^-(+\nu). \tag{1}$$

The negative charge of the emitted electron neutralizes the positive charge of the proton, so that the total charge of the products of the reaction is zero like the charge of the original neutron. The neutrino is needed to complete the energy and momentum balance. The parenthesis indicates that it is a particle whose presence is not observed experimentally, but only assumed, but we will spare ourselves this caution from now on.

Incidentally, we observe that all these reactions should always be denoted by the double sign, because it always happens, or there is always in principle the possibility, that one of these reactions may be opposed by the opposite reaction. Equation (1), for example, written in the most general form

$$e \rightleftarrows e' + \gamma$$

indicates the emission of a quantum of light when read in the left-to-right direction, according to the upper arrow, and the absorption of a photon when read from right to left, according to the lower arrow.

But there are others. We talked in the last lecture about the so-called μ meson. The μ meson is, as we have seen, the one of the two mesons that does not interact strongly with nuclear matter. Experimental observations, made with various techniques (Wilson's chamber, photographic plates, counters) show that when a μ meson stops far from the nuclei of matter, that is to say in vacuum (which happens notably for positive μ mesons that are kept away from nuclei by Coulomb repulsion), it soon spontaneously emits an electron, positive or negative depending on the sign of the charge of the meson itself, and endowed with great kinetic energy.

But, of course, this is not enough; and this is not enough for the simple reason that if the reaction consisted solely of the emission, by a stationary meson, of a fast electron it would violate the principle of conservation of momentum. Indeed, the initial momentum is zero (stationary meson), while the emitted electron possesses considerable momentum.

To set things right, it is necessary to admit that there is compensation for the momentum carried away by the electron; and this could be done by admitting that in the opposite direction a neutrino is emitted with the same momentum as the electron. But a more detailed analysis of the problem indicates that this is probably not enough. If the phenomenon consisted of the emission of an electron and a single neutrino, one would indeed have to expect that the kinetic energy of the emitted electron would always be the same, because the mass energy $mc^2$ of the original meson (which, incidentally, is worth about 100 million eV) should distribute itself in almost exactly equal amounts between the neutrino and the electron. The latter would then carry away $50 \times 10^6$ eV and the invisible neutrino would take away as much. Instead, the experimental results are different. Experiments definitely indicate that the emitted electrons, although having a maximum energy not far from 50 million eV, sometimes have lower energy: sometimes 30, sometimes 20, sometimes 40 MeV, assuming, with varying probabilities, all values between zero and 50 million eV. This indicates that it is not just one particle that we do not see and that serves to restore the momentum balance but that at least two particles are emitted.

In fact, in the speculations that are being made about the mechanism of this process (and these are speculations that are, at least for the moment, very hypothetical in nature), it is customary to admit that two invisible particles are emitted; and, for the sake of particle economy, we will assume that they are both neutrinos.

In conclusion, we may therefore indicate the reaction under consideration with the diagram:

$$\mu^{\pm} \rightarrow e^{\pm} + 2\nu. \tag{2}$$

There are some others. The other meson, the π meson, can also disintegrate, as is clearly seen in photographic plates. We saw some of them in the last lecture; there we can see the tracks of a π meson, which, having reached the end of its path, disintegrates, emitting, in any direction, a second meson which has considerable kinetic energy, and is recognizable, from the characteristics of the trace it leaves in the gelatin, for a meson of the μ type.

Here too, of course, the momentum balance must be safeguarded. Here too, in order to be economical in assumptions, and not to introduce more particles than is strictly necessary, we can admit that the particle emitted in the opposite direction, to compensate for momentum, is again a neutrino, And in this case the situation is different from the previous one, because experimentally it is observed that the μ meson is always emitted with the same energy. This indicates that in this case only one neutrino is presumably emitted.

We can therefore, if our speculations are correct, write the following reaction:

$$\pi \rightarrow \mu + \nu. \tag{3}$$

Of other reactions of the same kind we shall speak later, when the opportunity arises. We shall now deal with the point of view that is being followed today in attempts to construct a theory of phenomena of the kind I have described. These are, to tell the truth, mere attempts, and I would by no means say that they are satisfactory, like other branches of theoretical physics. But they are for the present the best that can be done and, pending some better idea, they constitute the method that is followed in the discussion of those phenomena.

Take any of the examples mentioned, for example, the disintegration of the $\pi$ meson, according to the reaction

$$\pi \rightleftarrows \mu + \nu$$

which we write with the double arrow, to indicate the possibility of the reverse reaction. The equation represents the possibility of a transition of the system from an initial state, in which there is a $\pi$ meson, which in general will be endowed with a certain velocity, and can thus be described by its momentum vector, and a final state, in which there are two particles, both different from the original one, and endowed with certain momentum, in directions generally different from that of the primary meson. If we represent the states by their corresponding momentum vectors (Fig. 1) we will thus have a transition between state A and state B.

**Fig.1**

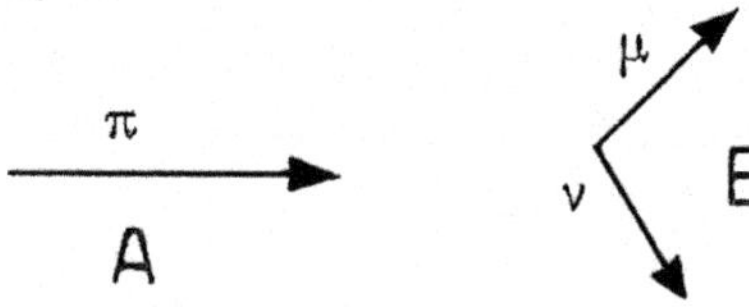


This description of the transformation in terms of the initial state and the final state of the system is actually quite incomplete. And it is so for two reasons: one of a secondary character which, with a little more analysis, could easily be eliminated, is due to the fact that the intrinsic quantities of motion of the particles have been neglected. In other words, we have considered them all as particles without spin. But this is a formal flaw that can be corrected without actual difficulty. A more serious omission consists of having neglected everything that takes place during the transformation. In other words, if we have a $\pi$ meson, traveling quietly in space, without interacting with anything, we can presumably attribute to it a determinate momentum; but when there are more than one particles present (and during the transformation, for a certain period of time, we must think of all three as present) and close to each other, there are naturally interactions; the particles mutually disturb each other and, to speak classically, their orbits will be curved and the momentum of each will not be constant.

It would therefore be necessary to try to describe also the phase in which the transition is taking place, in which therefore the state is a mixture of the initial state and the final state. An indication of the kind in Fig. 1 thus correctly represents only what takes place long before the transformation begins, and long after the reaction has taken place; but the really interesting phase, which is that during which the transformation, so to speak, is taking place, is very incompletely described.

In describing phenomena of this kind, quantum mechanics usually uses perturbation theory; and it uses perturbation theory simply as a fallback, because an exact theory of quantum phenomena is not possible with our mathematics, with few exceptions. Perturbation theory gives an approximate method that, at least sometimes, serves to give a fairly accurate idea of what is going on. In other cases, the approximation that can be achieved is rather unsatisfactory.

In any case, with the method of perturbations we obtain a formula that allows us to calculate the probability of transition from a given initial state of a system to a given final state. It turns out to be expressed as the probability that the process under consideration will occur per unit of time (or, if you like, as the number of processes that occur in the unit of time in a set of a large number of systems that are all in the same initial state), and therefore has the dimensions of the inverse of a time. Writing it in the form $1/\tau$, where $\tau$ is a time that is known to have the meaning of the "average life" of the system for the transition under consideration, we therefore have

$$\frac{1}{\tau} = \frac{2\pi}{\hbar} |\mathrm{H}|^2 n \tag{4}$$

where $\hbar$ represents, as is customary, Planck's constant h divided by $2\pi$; H is the matrix element corresponding to the transition from the initial state to the final state, whose explicit expression we will discuss shortly; and finally $n$ represents the "density of final states". In general, the density of final states means the number of final states contained in a unit energy interval, or, if you will, the number $\Delta\mathrm{N}$ of final states falling in an energy interval $\Delta E$ divided by the amplitude $\Delta E$ of the interval itself

$$n = \frac{\Delta\mathrm{N}}{\Delta\mathrm{E}} \cdot \tag{5}$$

It will be objected at once, at this point, that in the case we are examining the density $n$ of the final states turns out to be infinite; for, since the momenta of the two particles constituting the final state of our system are limited by the sole condition that their vector sum must be equal to the momentum of the initial particle, but otherwise are completely arbitrary, the final states constitute a continuum, and in each energy interval infinite ones fall.

But on the other hand, as can be seen from the calculation, the factor $|H|^2$ results in that case, and for the same reason, infinitesimal; and the product of the two factors converges to a finite limit. More precisely, the artifice usually used to perform the calculation of this limit is as follows: instead of considering the particles under consideration as free to move in an infinite space, one thinks of them as enclosed in a box of finite size and volume $\Omega$ (which is then made to tend to infinity in all directions); performing the calculation for this fictitious case, one finds that for any finite value of $\Omega$, the final states do not constitute a continuum, but a discrete succession, for which a finite density $n$ can be defined; correspondingly, the factor $|H|^2$ is finite. When we then let $\Omega$ to infinity, $n \rightarrow \infty$ while $|H^2| \rightarrow 0$ and the product $n|H|^2$, and thus the probability $1/\tau$, tends to a finite limit.

The next point is the calculation of matrix elements. Let therefore be in general

$$A + B \rightarrow C + D$$

a transition between initial particles A and B and final particles C and D (for simplicity we consider the case where there are two and two particles, but the reasoning for the general case of any number of particles would not be substantially different). According to the method of perturbations, the matrix element of the transition is represented by an integral

$$H = \iiiint \psi_C^* (C)\, \psi_D^*(D)\, V(A, B, C, D)\, \psi_A(A)\, \psi_B(B)\, d\tau_A\, d\tau_B\, d\tau_C\, d\tau_D \tag{6}$$

where $\psi$ (A) represents the wave function of particle A computed at the point occupied by the particle itself and which, for economy of symbols, we denote by the same letter used to denote the particle; and similarly $\psi$ (B, ...); the asterisk indicates the "complex conjugate" operation. V is a factor that depends on the nature of an interaction energy of the various particles present, which, at least in a certain order of approximation, will be able to be described as a mutual potential energy, a function of the coordinates of the four particles, V (A, B, C D); and the integral is to be extended to all the positions that the particles can assume, independently of each other. It is thus a fourfold volume integral.

In fact, the presence of the factor V (A, B, C, D), which represents the mutual positional energy between the four particles and which will presumably tend to zero when the particles move indefinitely away from each other, essentially implies the existence of an action at a distance between the various particles, which would interact even when they are not in contact (unless a field is introduced, which is the carrier of these interactions which would imply the existence of new particles, those arising from the quantization of the field itself; in other words, the interaction between the particles under consideration would lose the character of a fundamental interaction). Now there are serious theoretical difficulties-and I simply mention them without going into much detail—to admit the existence of actions at a distance. The difficulty consists essentially of the contradiction with the theory of relativity, because if two particles A and B interact at a distance, and particle A at a certain instant moves, the perturbation due to this movement should arrive instantaneously at the point occupied by particle B, located at a finite distance, instead of propagating with a speed which, according to the theory of relativity, must in any case be less than the speed of light.

To overcome this drawback, the function V that appears in the expression of the matrix element must choose itself in a rather singular way: that is to say, it is admitted to be non-zero (actually infinite) only when the four particles are in contact, that is to say, since we consider them as point-like, when the four points A, B, C, D coincide. This evidently allows a substantial simplification of the integral of formula (6): the four functions $\Psi_A$, $\Psi_B$, $\Psi_C$, $\Psi_D$ should all be calculated at the same point P, and the quadruple integral reduces to a single volume integral, extended to all positions of point P. Performing the calculation we find

$$H = f \int \psi_C^* (P)\, \psi_D^*(P)\, \psi_A(P)\, \psi_B(P)\, d\tau \tag{7}$$

where $f$ is an appropriate coefficient representing the intensity of the interaction between the four particles.

Let us mention only, at this point, that the expression we are deducing in this way for the transition matrix element depends essentially on the way one chooses to describe by wave functions the particles and their interactions: and of these ways one can choose many and very different ones from each other. First of all, the expression written for H contains only the wave functions themselves and not their derivatives: which, in many cases, would not be contrary to the relativistic invariance condition. In other cases, and precisely for all particles that have an intrinsic angular momentum, or, as we used to say, a "spin", the description by means of a single wave function is wholly inadequate, and in place of what I have referred to as simple scalar functions, multi-component quantities should actually be used. The best known example is that of the electromagnetic field, which can be described by means of a scalar potential and a vector potential: so that the electromagnetic wave is generally characterized by four quantities, the three components of the vector potential and the scalar potential. Similarly, particles endowed with spin 1/2, such as the electron, are represented, in Dirac's theory, by means of four-component functions. And so, in a large number of cases, in expression (7), instead of a product of four scalar functions, a quadrilinear combination of all the components relating to all the particles would appear, a combination that would contain many coefficients, instead of just one as in the simple case we examined. The coefficients would not be completely arbitrary, because they would be limited by the conditions of relativistic invariance. However, sufficient arbitrariness would remain for it not to be claimed that the value of the matrix element is substantially dependent on a single constant.

However, in many calculations in which only the order of magnitude is of interest, and sometimes also because one would not know any better, one uses simplified expressions of the (7) type, containing only one constant, while waiting to understand the phenomenon more deeply and to become able, a little at a time, to employ the formalism most suitable for dealing with the problem.

Already included in the expression we wrote for the matrix element is the necessity of conservation of momentum. Indeed, it is easily seen that this integral cancels out, if the conservation of momentum is not satisfied. If, for example, the state of particle A is one in which the particle has momentum $p_A$, the corresponding wave function contains a factor, dependent on the space coordinates, which is an imaginary exponential of the type

$$e^{i \frac{\vec{p}_A \, x \, \vec{r}_A}{\hbar}}$$

where the sign x represents the scalar product operation and $\vec{r}_A$ is the radius vector that locates the position of particle A. Therefore, the space-dependent part of the integrand of (7) contains four factors of that type, which can be combined into a single exponential:

$$e^{\frac{i}{\hbar} \vec{r} \, x \, (\vec{p}_{\mathrm{A}} + \vec{p}_{\mathrm{B}} - \vec{p}_{\mathrm{C}} - \vec{p}_{\mathrm{D}})}; \tag{8}$$

the momenta of particles C and D appear at the exponent with a minus sign because the conjugate complex quantities of the corresponding wave functions appeared in the integrand of (7).

If the momentum is conserved,

$$\vec{p}_{\mathrm{A}} + \vec{p}_{\mathrm{B}} = \vec{p}_{\mathrm{C}} + \vec{p}_{\mathrm{D}}$$

the exponent of (8) cancels out and the expression itself becomes constant and equal to 1.

Conversely, if momentum is not conserved, the exponential (8) comes to be an oscillating factor whose integral extended to all space is always zero: the matrix element is then also zero. This means that transitions can never occur between two states such that the final total momentum is different from the initial one. This property is also common to all other integrals of the same type that one encounters as one develops the theory more fully. They cancel if the conservation of momentum is not satisfied.

If the initial and final states are such that the conservation of momentum is satisfied, the exponential, as we have seen, reduces to 1 and the integrand becomes independent of position. The integrand then simply gives the volume of integration, i.e., the volume $\Omega$ of the box we alluded to in the beginning. In addition there will be a factor dependent on the normalization coefficients of the wave functions of the individual particles—and this is a point we need to dwell on for a moment, because the behavior of these coefficients turns out to be different, depending on whether they are particles obeying the Pauli exclusion principle, or particles obeying the Bose-Einstein statistic.

In the first case, the calculation of the normalization coefficient is very simple. The presence, in our volume box $\Omega$, of a particle of, for example, type A, will be represented by a wave function of the type mentioned above:

$$\psi_{\mathrm{A}} = b e^{\frac{i}{\hbar}\vec{p}\times\vec{r}}$$

where $b$ is the coefficient we want to determine. The normalization condition requires that the integral of the square of the modulus of the wave function, extended to the entire volume of the box, be equal to 1. But the square of the modulus of the exponential is equal to 1, and integration over the spatial coordinates simply gives the volume of the box $\Omega$. Therefore, we have

$$\int_{\Omega} |\psi_{\mathrm{A}}|^2 d\tau = b^2\Omega = 1$$

It is therefore simply

$$b = \frac{1}{\sqrt{\Omega}} \tag{9}$$

In the case that the particle obeys the Bose-Einstein statistic, the calculation is no longer so simple. Let us refer, for the sake of clarity, to a concrete and familiar case, which is that of the photon and the electro-magnetic field associated with it.

Suppose, then, that in the box of volume $\Omega$ there is a photon; for the sake of brevity we shall make a calculation purely on orders of magnitude, without caring about numerical factors close to unity, for which we shall put the right values in the final formulas. The box will thus be home to a system of electromagnetic waves, in which the electric field will have a certain value E, equal, if Gauss's system of unity is used, to the value of the magnetic field H. The energy density w is given, as is well known, by

$$w = \frac{1}{8\pi}(\mathrm{E}^2 + \mathrm{H}^2)$$

Recalling that E=H and disregarding the factor $1/4\pi$ we can say that the energy density is of the order of magnitude of $\mathrm{E}^2$, and the total energy enclosed in the box is of the order of magnitude of $\mathrm{E}^2\Omega$. This must equal the energy W of the photon; which gives us for the electric field strength, again as an order of magnitude

$$\mathrm{E} = \mathrm{H} \sim \sqrt{\frac{\mathrm{W}}{\Omega}}\,. \tag{10}$$

But in the case of an electromagnetic wave, what corresponds to the amplitude of the photon wave function is not the electric field, but the potential vector A, related to the electric field through the relation

$$\vec{\mathrm{E}} = -\frac{1}{c}\frac{\partial \vec{\mathrm{A}}}{\partial t}$$

Now, the time derivative of a periodically varying quantity is of the order of magnitude of the amplitude multiplied by the frequency $\nu$. We can therefore write, by multiplying numerator and denominator by Planck's constant $h$

$$E \sim \frac{h\nu}{hc}\mathrm{A} = \frac{\mathrm{WA}}{hc} \tag{11}$$

Comparing (10) with (11) shows that A, the analog of the wave function for the photon, is given, as an order of magnitude, by the relation

$$\mathrm{A} \sim \frac{hc}{\mathrm{W}}\sqrt{\frac{\mathrm{W}}{\Omega}}\,.$$

Coming back to the general case, and introducing the correct numerical factors, we similarly find that the absolute value of the amplitude of the wave function for a particle of total energy W obeying the Bose-Einstein statistic is

$$|\psi| = \frac{\hbar c}{\sqrt{2\Omega \mathrm{W}}} \tag{12}$$

and this expression is often used to calculate the matrix element related to a transition in which a particle of that type is involved.

In reality things are often even less simple, and one must proceed with great caution. In fact, it is not always true that the potential of a wave is the one to be used as the analog of the wave function in the calculation of the matrix element. If one considers a particle, such as, for example, the proton or better still the neutron, endowed with a magnetic moment, there is a direct interaction of the magnetic field with the magnetic moment of the particle and, to describe it, one would have to multiply the probability density of the neutron being at a certain position by its magnetic moment and by the magnetic field of the wave, which, as can be seen from (10), is proportional to $\mathrm{W}^{1/2}$, instead of to $\mathrm{W}^{-1/2}$ as the wave function in formula (12). This is a source of considerable lack of definition of the theory. If one knew how the "true" theory is to be made, one would naturally know in which cases a formula of the (12) type is to be used and in which cases a formula of the other type is to be used; but the "true" theory does not exist, for the time being. What can be said is that the justification of certain peculiar phenomena lies in this circumstance, and these phenomena have been discovered by theoretical physicists: for example the fact that the electromagnetic interactions of a meson are, at high energies, much stronger, and grow much more rapidly with energy if the meson has a high spin, than if the meson has a spin null or equal to 1/2.

Otherwise, at the present time, we know so little that the choice between a formula of the (12) type or a formula of the other type must unfortunately be made simply at random. And in the example we shall now describe we shall make use, for simplicity's sake, directly of formula (12).

Let us choose, as an example of the application of the general methods we have described, the simplest of the reactions we mentioned in the beginning: the conversion of a $\pi$ meson to a $\mu$ meson and a neutrino

$$\pi \rightarrow \mu + \nu \tag{13}$$

Of the three particles involved in the reaction it is believed (and even this statement must be made with some caution) that the $\pi$ meson obeys the Bose-Einstein statistics, while the $\mu$ meson and the neutrino would obey the Pauli principle. Formula (4) and the following then allow us to express the transition probability directly, by means of a formula (which turns out to be valid for both free and otherwise bound mesons) that contains only one empirical constant *f*, the one that appears in the expression of the matrix element H. It is appropriate to pause for a moment to show the analogy of this coefficient with electric charge. In a phenomenon of electromagnetic character the interaction can be represented essentially (and simplifying things somewhat) by the product of the charge density $\rho$ by the potential V, integrated over all space:

$$\int \rho V d\tau \tag{14}$$

Since this is an electromagnetic wave, we should have made use of the vector potential, which we did not do, as usual, for reasons of simplicity. On the other hand, the charge density can be written as the product of the charge of the electron by the square of the absolute value of the electron wave function itself

$$\rho = e\,|\psi|^2 \tag{15}$$

By introducing (15) into (14) we obtain, for the interaction energy

$$e \int |\psi|^2 V d\tau;$$

this represents, as is well known, the perturbation of the energy of a steady state of the electron due to the interaction of the electron with the potential V. In the case that we wish to deal with, however, with a problem of transition from one state to another, the rules of quantum mechanics teach us that the square absolute value of the wave function corresponding to the steady state should be substituted for the product of the wave function of the final state $\Psi_e'$ by the conjugate complex of the wave function $\Psi_e$ of the initial state; while the interaction potential V can be substituted for the wave function $\Psi\gamma$ of the photon that is emitted in the transition; the matrix element of the transition thus results in being represented by the expression

$$e \int \psi_e^* \psi_{e'} \psi_\gamma d\tau$$

$$\pi \rightarrow \mu + \nu.$$

similar to (7), except that the charge of the electron appears instead of the coefficient $f$.

This is thus presented as the coefficient that determines, so to speak, the intensity of the interaction between the electron and the electromagnetic field. If instead of an electron we had considered a proton, we would have obtained a similar result, and the coefficient would have been the same, since, as far as we know (although we cannot give the explanation) the electric charge of the proton is the same as that of the electron. In the case of the reaction $\pi \rightarrow \mu + \nu$. the matrix element is written as

$$\mathrm{H} = f \int_{\Omega} \psi_{\pi}^{*} \psi_{\mu} \psi_{\nu} d\tau \tag{16}$$

the coefficient $f$ thus has for the phenomenon under consideration a significance analogous to that possessed by the electric charge for electromagnetic phenomena and indeed even has the same dimensions as the charge. As for its numerical value, it cannot be established theoretically a priori but must be determined experimentally.

The integral (16) can be easily calculated from the considerations we made earlier. Since momentum must be conserved, integration of the factors dependent on spatial coordinates simply gives the volume $\Omega$ of the box; and for the absolute values of the wave functions we can use the expressions (12) and (9), taking into account the fact that the meson $\pi$ should be considered as a "Boson," i.e., as a particle obeying the Bose-Einstein statistic, while meson $\mu$ and neutrino obey the Pauli exclusion principle. In formulas

$$\mathrm{H} = f \int_{\Omega} \psi_{\pi}^{*} \psi_{\mu} \psi_{\nu} d\tau = f |\psi_{\pi}||\psi_{\mu}||\psi_{\nu}| \int_{\Omega} d\tau = f \Omega \frac{\hbar c}{\sqrt{2\Omega m_{\pi} c^2}} \cdot \frac{1}{\sqrt{\Omega}} \cdot \frac{1}{\sqrt{\Omega}} = \frac{f \hbar}{\sqrt{2\Omega m_{\pi}}} \tag{17}$$

where for the energy $W_{\pi}$ of the $\pi$ meson we put the rest energy $m_{\pi} c^2$ assuming that, in the initial state, the $\pi$ meson is at rest.

To calculate the transition probability we still need to find the expression of the density $n$ of the final states. As usual, we will simplify the calculation by admitting that, since the neutrino has a much smaller mass than the other two particles, and in particular the $\mu$ meson, we can neglect, in the final state, the momentum of the latter with respect to the momentum of the neutrino. Then we will certainly introduce in the final formula those modifications, easily understood, that result from a more rigorous calculation. Thus, calling the momentum of the neutrino $p_{\nu}$, the number dN of final states is obtained, as is well known, by dividing by the cube of Planck's constant the volume of phase space corresponding to a momentum between $p_{\nu}$ and $p_{\nu} + dp_{\nu}$ and to a position of the particle inside the box $\Omega$; that is to say, using $\hbar$ instead of $h$

$$d\mathrm{N} = \frac{4\pi p_{\nu}^2 dp_{\nu} \Omega}{8\pi^3 \hbar^3}$$

and the density n of states is obtained by dividing dN by the energy interval $dE = cdp\nu$

$$n = \frac{d\mathrm{N}}{d\mathrm{E}} = \frac{p_\nu^2 \Omega}{2\pi^2 \hbar^3 c} \tag{18}$$

We are now able to calculate the value of the transition probability $1/\tau$ and by introducing into formula (4) the expression of the density of final states given by (18) and the square of the matrix element given by (7). The volume $\Omega$ of the box, which we had been dragging along until now, cancels, which is comforting; and we get:

$$\frac{1}{\tau} = \frac{f^2}{2\pi\hbar^2 c\, m_\pi} p_\nu^2 \,.$$

Also taking into account the momentum of the meson that we had neglected, the result is slightly modified by the addition of a numerical factor, close to 1:

$$\frac{1}{\tau} = \frac{f^2}{2\pi\hbar^2 c\, m_\pi} \frac{p_\nu^2}{1 + \frac{v_\mu}{c}}$$

where $v_\mu$ , is the speed at which the $\mu$ meson is emitted, which is quite small compared with the speed $c$ of light.

In the second member of the expression now written everything is known except the constant $f$. By experimentally measuring the average lifetime $\tau$ of the $\pi$ meson for the process under consideration, we can then determine the value of $f$, which, as we have mentioned, is the size of an electric charge; the value of f thus obtained is much smaller than the charge of the electron, and it is therefore convenient to express it as a fraction of this charge. To be precise we obtain

$$f \sim \frac{e}{4 \cdot 10^5} = 2,5\ 10^{-6}\, e.$$

The intensity of the interaction that causes this transition is thus about two and a half millionths of the intensity of the interaction between the electron and the electromagnetic field.

To conclude, I wish to mention a most remarkable circumstance, which arises in the study of these phenomena. A calculation analogous to the preceding one can be developed for many other reactions; and in all reactions in which it is essentially the electric charge that determines the interaction the same coefficient $e$ appears, because all known elementary particles, although essentially different in all their other characteristics, when they are charged (such as the proton, the meson, the electron) always have the same charge, positive or negative, but invariable in absolute value: a fact for which, as we have already mentioned, no satisfactory explanation can be given (although some hypotheses have been put forward), but which is experimentally well established.

Now, there are three reactions of a non electromagnetic character, two of which we have already discussed in the beginning, in which the interaction constant has the same dimensions (which, however, do not coincide with those of an electric charge, being the product of a volume by a work). The three reactions are schematically as follows.

$$\begin{cases} \mathrm{N} \to \mathrm{P} + e + \nu \\ \mu^{\pm} \to e^{\pm} + 2\nu \\ \mathrm{P} + \mu^{-} \to N + \nu. \end{cases} \tag{19}$$

The first is the $\beta$ transition of nuclei; the second is the disintegration of the $\mu$ meson, with average lifetime $\tau = 2.1 \times 10^{-6}$ s. The third, which we have not yet discussed and will mention without giving much detail, has been highlighted by the experiment of Conversi, Pancini, and Piccioni on the capture, in matter, of $\mu$ mesons: a negative $\mu$ meson would be absorbed by a proton, which would turn into a neutron, with emission of a neutrino. Of course, even in this case, the presence of the neutrino is purely hypothetical; however, from the study of the results of the experiments mentioned above, data are derived that allow us to calculate the interaction constant in this case as well.

Thus we have three interaction constants, $f_1$, $f_2$, $f_3$ having the same dimensions and relating to three reactions which a priori, as far as we know at the present time, are completely different from each other. Even assuming without any doubt the validity of the schemes indicated (which, however, as we have already seen, must be taken with some caution), it can be observed, for example, that in the second reaction two neutrinos are emitted, while in the first and third only one is emitted.

The very remarkable, and very suggestive fact, to which I have alluded, and for which there is no explanation, is that the numerical values of these constants, calculated by the methods previously indicated, turn out to be very close to each other. It cannot be said that they are exactly equal but they are within a range of about 10 units: between $10^{-48}$ and $10^{-49}$. It would not be easy to say more, because, as we have repeatedly mentioned, our theoretical way of seeing these phenomena is still extremely rudimentary, and we can consider ourselves lucky if the formulas correctly represent orders of magnitude. But, as orders of magnitude, these quantities are certainly very close, especially if we note that, after all, since they are such unusual numbers, we could consider the order of magnitude of their logarithms, which are $-48$, $-49$, respectively.

This fact has been noted independently by various researchers: among others Wheeler and Puppi have pointed out the strange numerical concordance of some of these constants and observed that this would suggest (if one wishes to hazard such an interpretation), the existence of a property analogous to electric charge, which occurs in the same way in all elementary particles. We have already noted several times that even in the electromagnetic case we do not know the reason, which is certainly extremely significant, why particles so different as the electron, the proton, the meson, all possess the same electric charge. In that case our knowledge of electromagnetic phenomena is such that we can assert that the electric charge always has, to a very high approximation, the same value, and not simply the same order of magnitude. In the case of reactions of the (19) type our knowledge is much more limited, and we cannot say more than what we have seen.

It may be, of course, that with a more complete analysis, when it is possible to make it, those phenomena appear similar to each other but the values of the constants do not turn out to be exactly the same; in which case that concordance would be merely a chance, perhaps a little strange but not impossible. But if, on the day when a rigorous and not merely grossly approximate theory can be made, it is found that the analogous magnitude f of the electric charge is indeed the same for those three phenomena, so different from each other, this would certainly be an indication of very great theoretical significance and probably rich in consequences-although at the present time these consequences cannot be foreseen.

## Biographical Notices on Lucio Mezzetti

### AT and PT

Lucio Mezzetti was born in Padua on November 28, 1919. From August 1944 to October, 1945 he was substitute assistant professor in place of Ettore Pancini (1915–1981)—then assistant on leave for recall to the army—at the Institute of Physics in Rome. From November 1945 to February 1949 he was enrolled as tenured assistant professor in the Physics Institute of Rome. On 1 March 1949 he became Assistant Professor of Advanced Physics and confirmed in that role until 1958–59. He participated in cosmic ray experiments at the Testa Grigia laboratory: in 1956-57, together with Ruggero Querzoli (1923–1989) he carried out research on penetrating showers produced in lead and carbon with the technique of Geiger and Müller counters in coincidence. From 1964 to 1967 he was director of the Laboratori Nazionali Frascati (LNF) (National Laboratories Frascati) where he participated, from March 1961, in the design of experiments with ADONE together, among others, with Bruno Touschek (1921–1978), leader of ADA (1961). In 1961 he became Professor of the University of Trieste and later in Padua.

He died in 2002.

For further informations see: Bernardini (2005); Bonolis et al. (2021); De Maria and Orlando (2008); Gigli Berzolari (2005); La Rana and Rossi (n.d.); Sestili (2013/2014).

## References

Bernardini, Carlo (2005). "AdA: The First Electron-Positron Collider" Physics in Perspective 2004 6: 156-183.

Bonolis, Luisa; Bossi, Fabio and Pancheri, Giulia (2021). "The Frascati National Laboratories. Past and Future" Il Nuovo Saggiatore 2021 37 (3-4): 47-60.

De Maria, Michelangelo; Orlando, Lucia (2008). Italy in space1957-1975 (Paris: Beauchense 2008).

Gigli Berzolari, Alberto (2005). "Roma: con Carlo negli anni del dopoguerra" Il Nuovo Saggiatore 2005 3-4: 27-30. In the Section "Il Nostro Mondo", In Memoria di Carlo Castagnoli.

La Rana, Adele; Rossi, Paolo (eds) (to be published). "Preparatory material for 'I Fisici Italiani' Vol. 3". Private communication.

Sestili, Giorgio (2013/2014). ADONE. Storia dell'anello di accumulazione per elettroni e positroni Tesi del Corso di Laurea Specialistica in Fisica, A.A. 2013/2014, Relatore Giovanni Battimelli. https://www.giorgiosestili.it/wp-content/uploads/2022/12/Touschek_ADONE_Storia_dellanello_di_accumulazione_per_elettroni_e_positroni.pdf accessed in September 27, 2023.

# Third Lecture Theories on the Origins of the Elements

**Enrico Fermi**

## Edited by E. Pancini

All known matter consists of various chemical elements, each present with different abundance, so that the problem arises, experimental at first, theoretical later, of understanding why some elements are abundant and others rare.

The problem is first and foremost an experimental problem and, without wishing to discuss the matter here in detail, a few general observations are sufficient to understand why. What is sought to be established is the quantity of the various chemical elements found, shall we say, throughout the Universe, or at least in a large portion of it, and, of course, the result that can be expected to be obtained depends largely on the samples of matter chosen for analysis. For example, if one can determine the relative amounts of oxygen, iron, hydrogen and other elements found in the part of the Earth's crust that is accessible to our direct observations one will derive, for each of them, a certain relative abundance; but if one determines instead, for example, the percentage of the same elements. By analyzing meteorites one will find a distribution of elements different from that found in the surface rocks of the Earth.

The problem, therefore, is essentially a problem of choosing specimens for analysis rather than a problem of chemical analysis.

(*) Held October 7th, 1949 in the lecture hall of the Physics Institute of the University of Rome.

E. Fermi (✉)
University of Chicago, Chicago, IL, USA



A. Treves and P. Tucci (eds.), *Enrico Fermi, Atomic Physics Lectures*, History of Physics,
https://doi.org/10.1007/978-3-031-68690-0_6

The question of course is not new. The data that will be presented here are obtained from rather recent research by Harrison S. Brown[1] of the University of Chicago who extended, expanded and refined Goldschmidt's results.[2] The data were obtained by analyzing a large number of specimens, and this is a guarantee of their reliability because data obtained from a particular specimen reveal rather the special characteristics of that specimen than what may be considered the cosmic distribution of the elements.

Remarkably, in spite of this observation, the conclusion of these analyses is that, when appropriate care is taken in the choice of samples to be analyzed, the results are extremely uniform even when drawn from materials of very different origin. It is possible, for example, in some favorable cases, to assign the ratio of the cosmic abundance of two elements with an accuracy of the order of 1 or 2%.

It should be noted that measurements of the abundance of elements made on the Earth's crust, while being of very great practical importance, have a rather more limited theoretical importance because the Earth's crust, indeed all the material of which the Earth is composed, has been subjected during geological eras to a profound chemical separation so that meaningful results could only be obtained from an analysis made by studying samples taken from areas ranging from the Earth's surface to its center, which of course is impossible.

Fortunately, this impossibility of taking samples from the inner part of the Earth can be turned around by examining the constitution of meteorites, which experts say are like specimens taken from various parts of vanished planets. Thus if the Earth, by a cosmic catastrophe, were to fall apart, meteorites from its crust would be meteorites consisting essentially of stone, while those from its central part would consist essentially of iron, or more precisely, of an alloy consisting, in large part, of iron and then of nickel and then in smaller and smaller quantities of other elements.

Indeed, a statistical analysis of meteorites arriving on Earth (and the meteorites arriving on Earth are precisely of these two types) indicates that the ratio of the amounts of matter in the form of stone meteorites to the amounts of matter in the form of iron meteorites is not very dissimilar to the ratio of the stony to the ferrous part of the Earth as revealed by investigations made at depth by means of seismic waves.

The research we speak of has been done largely by patiently collecting a large quantity of meteorites and carrying out extremely accurate quantitative analyses on them. It should be noted that the problem of performing these analyses is much less simple than might seem because many of the elements, indeed, as we shall see, the majority of the elements, are so rare as to be present in quantities of the order of a few parts per million and even less. One of the greatest difficulties of the problem is therefore to find a way to perform quantitative chemical analyses of extreme delicacy.

To overcome this difficulty it has even been necessary (at least in the recent studies that have been carried out at the University of Chicago) to make use of atomic piles

[1] "Rev. Mod. Phys.," 21, 625 (1949).

[2] V. M. GOLDSCHMIDT, "Geochemische Verteilungsgesetze der Elemente und der Atomarten," IX, Oslo 1938.

that have been used to irradiate the material under examination in order to be able then to observe the characteristic activities of the elements one is interested in. The recognition of elements is done, therefore, by this route essentially by radioactive rather than chemical methods. By these means it has been possible to analyze the material arriving on Earth in the form of ferrous and stony meteorites, and by taking averages on the data obtained with appropriate criteria, a table has been constructed which, for the majority of the elements, coincides very well with other data of absolutely different origin such as, for example, those obtained from spectroscopic observation of the stellar atmosphere. Thus we have an indication that the matter of which meteorites are composed is not substantially different from that which constitutes stellar atmospheres.

There are actually some very notable exceptions, but they are easy to explain. For example, there are some elements that are practically missing in meteorites, or are present in much smaller quantities than they should be. This is the case of the noble gases which are, both in the Earth and in meteorites, present in much smaller quantity with regards to their cosmic abundance because, in the process of the formation of the planets, they have not been retained in them. Another even more notable exception is that of hydrogen, but this exception, too, is explained by a reasoning similar to that made for rare gases.

Facts of this kind make it clear that the results obtained from the analysis of meteorites (from which most of the quantitative data on the cosmic abundance of the elements are drawn) must be reworked through a careful chemical discussion. This, by the way, is unfortunately almost entirely arbitrary since it involves assumptions about the mode of formation of the meteorites themselves and the chemical-physical characteristics of the environment in which they passed their history. In short, the analysis of the meteorites, element by element, must be supplemented through chemical-theoretical considerations. These will make it possible to judge whether or not it is to be believed that the element in question has preserved its cosmic proportions in the meteorites.

In addition to meteorites, stellar atmospheres (through spectroscopic analysis) and, to some extent, the clouds of matter in interstellar space through analysis of their absorption spectra have been studied. The latter data, however, are extremely scarce and to be used only as subsidiary. What is, however, very remarkable is the fact that all these data (where due account is taken of the exceptions mentioned, which can, moreover, be justified by very persuasive arguments) drawn from the analysis of celestial objects as diverse as stellar atmosphere, interstellar matter or meteorites, coincide very well. They vary considerably, on the other hand, as already mentioned, in the data that can be derived from the analysis of the Earth's crust since, we repeat, the Earth's crust is not a faithful sample of what can be considered to be the material of which the Earth is made.

That said, it is interesting to examine the table given here in which numbers are given representing the relative abundance of various elements (Table 1).

These data are taken from the work of Harrison S. Brown, which may be considered the most up-to-date. The numbers listed in the table refer to some of the most significant elements and are sufficient to highlight some very strange features

**Table 1** Relative Abundance of the Elements

| | | A | Z | Atoms per $10^4$ atoms of silicon |
|---|---|---|---|---|
| H | .......... | 1,01 | 1 | 3,5 x $10^8$ |
| He | .......... | 4 | 2 | 3,5 x $10^7$ |
| Be | .......... | 9,02 | 4 | 2 x $10^{-1}$ |
| C | .......... | 12,01 | 6 | 8 x $10^4$ |
| O | .......... | 16,00 | 8 | $10^5$ |
| Si | .......... | 28,06 | 14 | $10^4$ |
| Cl | .......... | 35,46 | 17 | 2,5 x $10^2$ |
| Mn | .......... | 54,93 | 25 | $10^2$ |
| Fe | .......... | 55,85 | 26 | 2,6 x $10^4$ |
| Co | .......... | 58,94 | 27 | 1,6 x $10^2$ |
| Ni | .......... | 58,69 | 28 | 2,0 x $10^3$ |
| Cu | .......... | 63,57 | 29 | 7 |
| Ga | .......... | 69,72 | 31 | 5 x $10^{-3}$ |
| Sr | .......... | 87,63 | 38 | $10^{-1}$ |
| Cd | .......... | 112,41 | 48 | 2 x $10^{-2}$ |
| Cs | .......... | 132,91 | 55 | $10^{-2}$ |
| Pt | .......... | 195,23 | 78 | $10^{-1}$ |
| Pb | .......... | 207,21 | 82 | 4 x $10^{-3}$ |
| Th | .......... | 231,12 | 90 | $10^{-2}$ |
| U | .......... | 238,07 | 92 | 3 x $10^{-3}$ |

of the trend, as a function of atomic number, in the relative abundance of the various elements. They represent the number of atoms of each element present, on average, in cosmic matter for every $10^4$ atoms of silicon 14.

In analyzing this table it is convenient to begin with hydrogen, which is not only the simplest among the elements but also the most abundant: the number of its atoms present in cosmic matter for every $10^4$ silicon atoms is three or four hundred million. Both in the periodic system of elements and in the scale of abundances after hydrogen comes helium, the relative abundance of which is, on our scale, 35 million. For the elements that follow helium the abundance drops very rapidly to extremely low values: lithium, beryllium and boron are extremely rare. The relative abundance of beryllium, for example, is 2 tenths (i.e. for every atom of beryllium there are on average 50,000 of silicon). As can be seen, between helium and beryllium there is a jump by a factor of the order of 100 million.

The other light elements that follow the three now mentioned in the periodic system have abundances little different from that of silicon. To carbon, for example, is to be ascribed the abundance $2 \times 10^4$. Immediately afterwards oxygen marks an upward point: $10^5$. It is, after hydrogen and helium, the most abundant element by number of atoms (but not by weight).

Proceeding on this scale one finds abundances of the order of a few units until one reaches iron to which is allotted a remarkably high abundance: $2.6 \times 10^4$. Then cobalt: $2.6 \times 10^2$, nickel: $2 \times 10^3$, and, continuing in the order of the periodic system, at this point the abundance begins to decline rapidly and does not rise again. From gallium on up to uranium, abundances are found that fluctuate more or less irregularly between a tenth and a hundredth. A slight exception is lead, which has a somewhat higher abundance, but the amount of lead may be thought to have increased as a result of the decay of radioactive substances immediately above it. Another exception, in the opposite direction, is uranium but it can be thought of as being depleted due to its radioactive decay.

All this talk will probably become more evident if one represents in a diagram (Fig. 1) the relative abundance of the elements as a function of atomic number. From this diagram one can see how immediately after the extreme point represented by hydrogen and helium there is a tendency to show, albeit with a very great irregularity, a downward trend in the relative abundance of the elements. So that anyone who dared to draw a curve through these points, and did not take too much account of the irregularities, could draw the curve shown in Fig. 1. And, whoever wished to believe that, exceptions being justified, this curve represents with good approximation the relative abundance of the elements in cosmic matter, would have to conclude that the relative abundance of each element is as essential a characteristic of it as is, for example, its atomic number, or its energy of formation, or its mass. There is thus the impression that the relative abundance of each element is really a property of it, connected, as a matter of fact, both with the other properties of the element and with the mechanism, otherwise unknown, by which the element was formed.

**Fig.1**

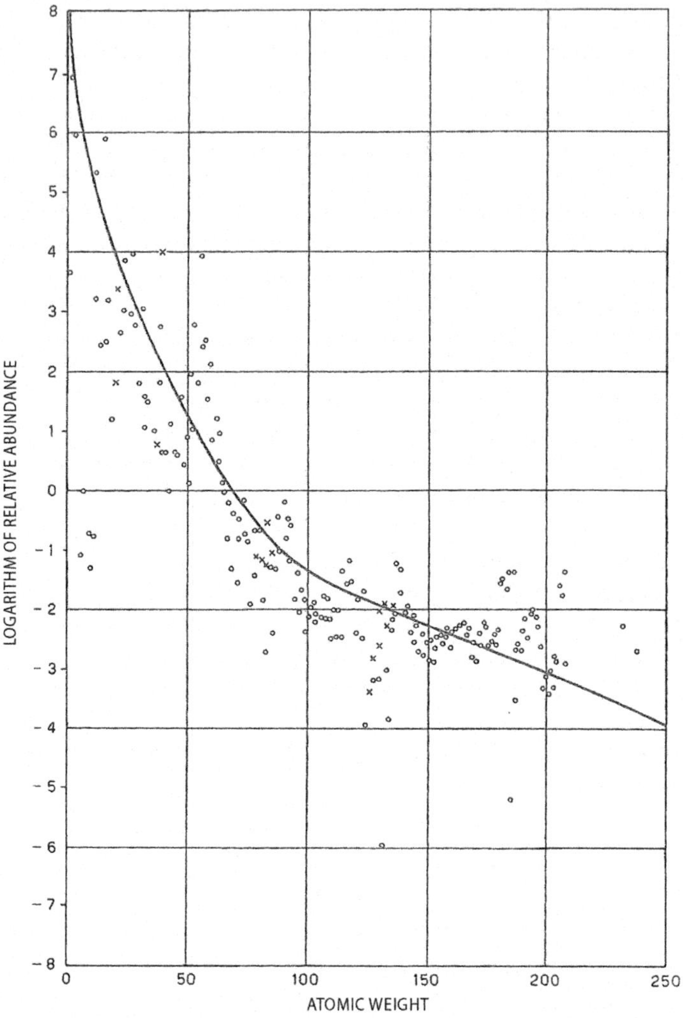


Of course, in such a discussion one must also take into account the abundance of the individual isotopes of each element, but this is not a complication of the problem since the relative abundances of the isotopes of each individual element are known and normally quite constant. Thus when the abundance of each element is known it is simple arithmetic to calculate the abundance of the isotopes.

Even in studying the relative abundances of isotopes regularities will be noted to which it is worth drawing attention since there will be occasion to return to them later.

A diagram of the isotopes is given in Fig. 2, indicating in the abscissae the number of protons and in the ordinates the number of neutrons that make up each nucleus. As can be seen, the various elements are distributed in an area that initially has the direction of the bisector of the axes and then bends upward (which means that in nuclei of small atomic number the number of protons is equal to that of neutrons while for high atomic numbers the percentage of neutrons is, increasingly, more than 50%). Now it is found that almost invariably, in the lower part of the periodic system—that is to say for light nuclei—the most abundant isotopes are richer in protons or, which is the same, poorer in neutrons. Then there comes a transition zone, and finally in the upper part an entirely opposite trend is observed. The most abundant isotopes tend to have more neutrons than protons.

**Fig.2**

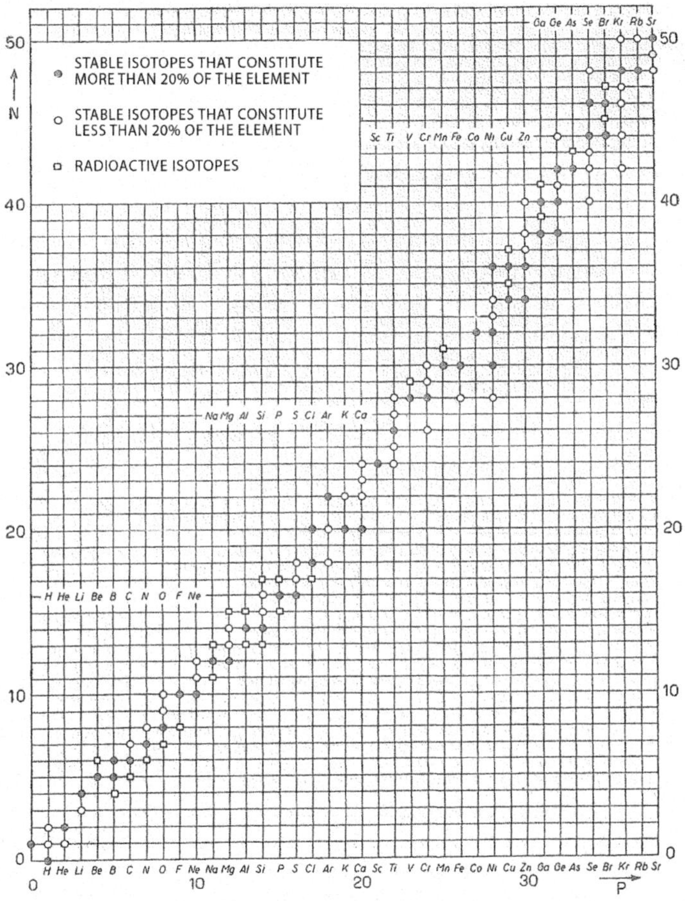


Of course, the idea of justifying all these facts, that is to say, of justifying the abundance of the individual elements and, for each element, the abundance of its isotopes, is, of course, an extremely ambitious program and we are, no doubt, a long way off from the solution of this problem. However, there have recently been attempts in this direction which lead to results that are far from satisfactory. This fact does not mean that they are not extremely interesting in that they represent an attempt at research in a direction that will most likely be one of the most important in the future. On the other hand, it is obvious that if the solutions reached today are not satisfactory, it cannot be ruled out that conclusive steps may be taken on this road later on.

One of the most natural hypotheses that has been put forward, and has in fact been put forward by various people for various lengths of time, is that the elements that we find in nature are the result of a process that is based on a kind of chemical equilibrium or, as they say, superchemical equilibrium. In other words, one may wonder whether it is permissible to imagine that by placing the constituent elements of the elements, i.e., protons and neutrons, in a cauldron and then heating it to a suitable temperature and then, when this matter is, so to speak, cooked just right, by suddenly cooling it down one can obtain a mixture of elements that resembles what seems to us to constitute the Universe.

Many attempts in this direction have been made but the results obtained are, in truth, not very convincing. Of course, the temperatures and pressures that are housed in this sort of cauldron must be imagined to have rather astonishing values if results are to be obtained that do not absolutely diverge from the experimental ones. The temperature, for example, should be around 10 billion degrees and the pressure around one million grams per $cm^2$. The necessity of such temperatures and pressures can be understood without difficulty as soon as one considers that the temperature must be very great in order to give rise, conspicuously, to nuclear reactions and that the pressure must be very high for there to be a possibility of forming very heavy nuclei. In fact, if, at such high temperatures, the pressure were not proportionately high, all nuclei consisting of many particles would disintegrate and there would, therefore, be no possibility of existence for the heavy nuclei that are instead found in nature.

It remains a fact, moreover, that starting from such assumptions one cannot obtain a distribution of the elements that bears much resemblance to the real one. For example, the relative abundances of the various isotopes, turn out to be distributed completely at random and without the slightest resemblance to those observed experimentally.

The more modern theories are based, however, on a somewhat different scheme. We shall confine ourselves here to speaking of only one of them, the one which, in my opinion, is the most interesting although it cannot be considered, in any way, capable of giving a satisfactory explanation of the facts. It is mainly due to Gamow who, being as everyone knows a prankster, associated himself with two other physicists, Alpher and Bethe, for the purpose, perhaps, of playing on the fact that, the three names, Alpher, Bethe and Gamow, read in the American way by mispronouncing the words a little, may sound like the names of the first three letters of the Greek alphabet. In reality, however, the substantial contribution made to the theory we are dealing with is, essentially, that of Gamow and, in part, that of Alpher. Bethe, on the other hand, seems to have been involved only to complete the wordplay.

Be that as it may, the theory can be divided, essentially, into two parts. The first has for its foundation the observation, which is actually not new, of the possibility of forming elements, even when temperature and pressure do not assume such staggering values as those mentioned just now, as long as we assume that we form the elements through successive additions of neutrons. Without dwelling, for now, on explaining the origin of these neutrons let us try to give an idea of the way in which this formation can take place.

Let us refer again to an N P diagram (Fig. 2) in which each element is presented by a point whose abscissa is equal to the number of protons and whose ordinate is equal to the number of neutrons it contains. As noted above, the stable elements are all in a very well-defined zone.

If we suppose, now, to expose a certain element to a "bath" of neutrons it may happen that its nuclei will capture one of these neutrons. Thus, if the composition of this nucleus is represented by point A in Fig. 3, after capture the newly formed nucleus will have a composition represented by point B, which is obtained by taking a step upward from A (in fact N increased by one and Z remained constant).

**Fig.3**

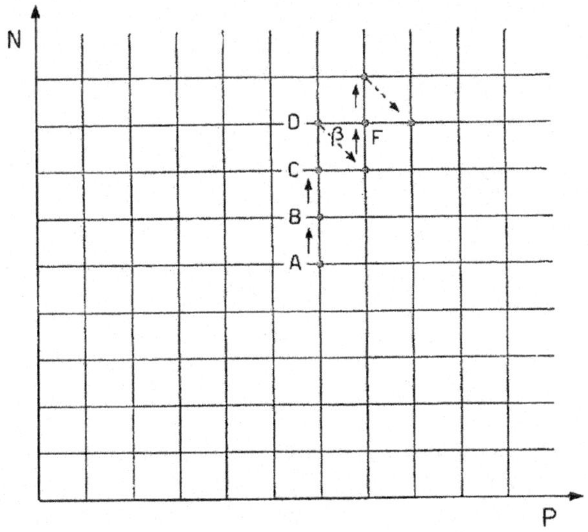


The new nucleus will be able, in turn, to absorb another neutron giving rise to an element represented by the C point and so on, until it ends up leaving the zone of stable elements. The unstable element formed will evidently be beta radioactive and will therefore disintegrate by a beta process that is a change of a neutron to a proton. The new representative point will, therefore, be obtained by taking one step downward (decrease of a neutron) and one step to the right (increase of a proton).

If, now, there are still neutrons present the nucleus thus formed will be able to absorb another neutron, then yet another neutron, then emit a beta beam. And so little by little it will climb up the slope of stable elements. Very heavy elements are thus gradually being formed through a mechanism of successive additions of neutrons to light nuclei that are assumed to be pre-existing.

At this point, to be ambitious (and as we shall see Gamow has even more ambitious claims than these), one can even propose to explain the formation of all elements from neutrons alone.

Suppose, in fact, that in a region of space, at a certain instant, neutrons are present. Now, the neutron, as is well known, is not a stable particle, indeed its average lifetime is rather short (it has not yet been measured very well but it cannot be very different from 15') and, therefore, after about ten minutes, half of the neutrons will have decayed giving rise to as many protons. But neutrons and protons have a certain affinity and neutrons tend to attach themselves to protons, thus forming deuterium nuclei. By this route, starting initially from neutrons alone, by decay of them and association with the protons generated, the first light nuclei can be formed, and then from these, by a process of the kind described just now, the formation of the heavy elements can presumably be arrived at.

**Fig.4**

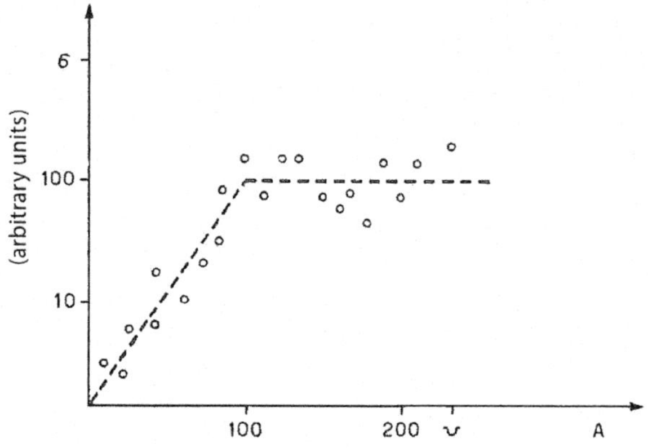


Gamow made an attempt to investigate this model (or rather a model similar to it) from a quantitative point of view. Of course, for a quantitative investigation it is necessary to introduce data on the probability of neutron capture by a certain element, because that is what essentially determines the rapidity of the process of this phenomenon. Now, there is a lot of data on slow neutron capture, but presumably phenomena of this kind occurred at a high enough temperature for it to be more convenient to take—and Gamow took—data corresponding to the capture of rather fast neutrons. Figure 4 shows the cross sections for fast neutron capture as a function of atomic weight.

By simplifying, and perhaps oversimplifying, what the experimental results really gave, Gamow assumed that these neutron-capture cross sections grew linearly for values of atomic weight between 0 and 100 and then remained constant as is indicated by the dashed line drawn in Fig. 4. Now, if one looks at the figure and takes into account that the scale of ordinates is logarithmic, one can judge how strong Gamow's schematization is. However, before criticizing it, it is worth following Gamow's reasoning to the end.

Assuming then, for the moment, that the cross sections are indeed those required by Gamow one can write very simply the differential equations describing how the subsequent formation of heavier and heavier elements takes place. Let us call $N_a$ the number of atoms of atomic weight $a$; the derivative of this number with respect to time will depend on two terms. One represents the growth in the number of atoms of weight *a* due to the aggregation of neutrons to atoms of weight *a*–1 (and this will be a positive term proportional to $N_{a-1}$ , to the cross section $\sigma_{a-1}$ of element *a*-1 and proportional to the flux, $\Phi(t)$, of neutrons). Then there will be a negative term that similarly represents the decrease in $N_a$ due to neutron absorption that transforms *a* atoms into *a*+1 atoms. In formula

$$\frac{dN_a}{dt} = \Phi(t)(\sigma_{a-1} - N_{a-1} - \sigma N_a)$$

$$(a = 1,\ 2,\ \ldots 238).$$

Equations like this will have to be written for each value of a, and a system will be obtained that can be solved, once the neutron flux is fixed, by deriving the way in which the abundances of individual elements evolve over time.

**Fig.5**

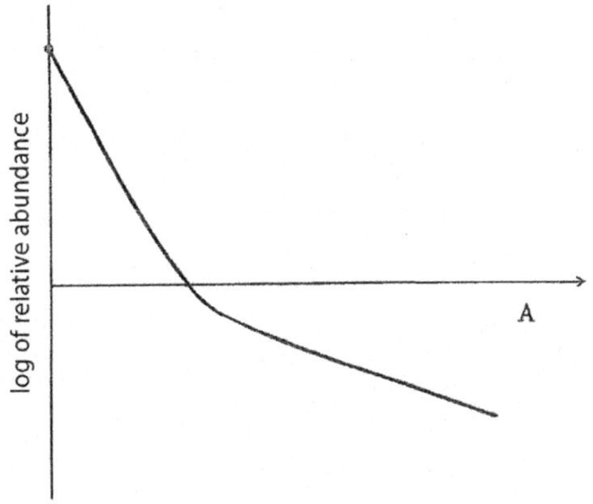


Now, the most significant result (and which would be even more significant than perhaps it is if the curve of cross sections chosen by Gamow were a more faithful representation of the experimental facts) is that, because of the singularity of this curve—that is to say, of the fact that for a certain atomic weight the tendency for the cross sections to increase comes to an abrupt halt—it happens that, choosing suitable time and neutron flux, one finds a distribution of element abundances of the type depicted in Fig. 5 and which is not very dissimilar to the experimental one (Fig. 1).

Of course, the result we get depends on the time chosen in the sense that, if we fix a certain neutron flux, the material must be exposed to its action for a suitable time—in fact, if the time is too long, too many heavy elements would be formed, if it is too short too few. But, by "cooking" the material as it were to the right degree, one is able to obtain something that bears some resemblance to the experimental data. This resemblance goes so far as to give, for elements of high atomic number, an isotope distribution similar to the real one. In the area of the light elements the result is, on the other hand, contrary to the experimental one but it may be thought that subsequent heat treatment, at even not excessively high temperatures, may have changed the situation.

Gamow, as noted above, was not satisfied with these results and tried to take a further step, a very risky and almost certainly erroneous step. Almost certainly as erroneous as what one does when, to explain facts, one poses very precise hypotheses because, as is obvious, the more precise the hypotheses the easier it is to prove them wrong. Gamow, however, proposed to determine the time in which the formation of the elements described above occurred by resorting to the theory of the expansion of the Universe. This theory is connected with that of general relativity and we will try to give a hint of it here.

**Fig.6**

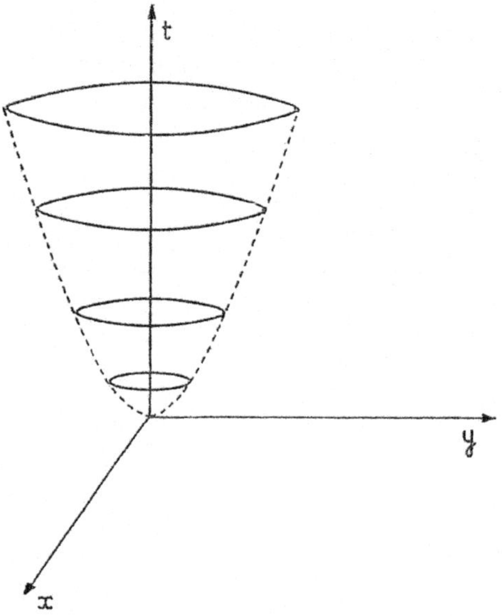


Unfortunately, even for general relativity, as for other physical theories, there is no single theory, and this always implies a certain freedom of choice, a choice that at the present time cannot be made on the sure basis of experimental indications. But if we rely on the simplest of the theories of relativity, the one in which the so-called cosmological terms are not taken into account, a theory of the expansion of the Universe can be constructed, and has been constructed, along the following general lines.

It starts with the assumption—which if nothing else has the merit of being very simple—that energy density (matter and radiation) is uniform throughout the Universe, at least when averages are taken over very large regions of it.

It is also admitted that space has constant curvature which is the same as saying that the Universe is required to be homogeneous not only in terms of energy density but also in terms of its geometrical properties. From this assumption it follows that the shape that the Universe has at a certain determined time must either be that of a sphere or that of a pseudo-sphere. For special reasons connected with the present density of matter one must choose the pseudo-sphere, which is a sphere with an imaginary radius that cannot be represented, in any way, in a figure. But if for the moment we disregard the fact that the object we want to talk about is a pseudo-sphere and not a sphere, and moreover we limit ourselves to representing only three of its four dimensions (time and two spatial coordinates) it will be possible to give (Fig. 6) an idea of how the Universe evolves by expanding. In Fig. 6 the Universe is represented by circles whose radius increases with the passage of time.

If we now do the calculations with the formulas of general relativity, we can find a relationship, which links the speed with which it expands, the radius, $r=iu$, of the pseudo-sphere with the energy density $w$:

$$\left(\frac{du}{dt}\right)^2 = c^2 + \frac{Kc^2}{3}u^2 w. \tag{2}$$

This formula says that the square of the derivative with respect to time of the modulus of the radius u of the pseudo-sphere is equal to the square of the speed of light plus a term containing the radius itself, the energy density w, and a constant K that is related to the universal gravitation constant G in the following way

$$K = 8\pi\frac{G}{c^4}$$

and has a value of about $2 \times 10^{-48}$ $\text{dyn}^{-1}$. Now, if one wishes to use this formula to describe the expansion of the Universe when its radius is very small, one can see that the first term of the second member becomes negligible since the energy density grows much faster than the square of the radius. Formula (2) is then simplified as follows:

$$\left(\frac{du}{dt}\right)^2 = \frac{Kc^2}{3}u^2 w. \tag{3}$$

At this point Gamow made a very interesting observation: if one admits that, in cases where formula 3 is usable, energy is essentially radiation energy, one can, using the formula itself, derive a relationship between temperature and time that contains no constants other than universal constants. One comes, in this way, to eliminate the arbitrariness as to the value of the temperature at which the nuclear reactions that gave rise to the elements would have taken place, an arbitrariness, playing upon which, one could have obtained, within certain limits, any result one desired. We give here, without demonstration, the formula that links temperature to the time elapsed from the instant in which the Universe had infinitesimal dimensions.

$$T = \left(\frac{3}{4Kc^2\sigma}\right)^{1/4} \frac{1}{\sqrt{t}} \quad \text{degrees.} \tag{4}$$

In it $\sigma$ represents the Stefan's law constant, T the absolute temperature and $t$ the time. As has already been warned it is a formula, approximate and valid for values of $t$ that are not too large, for example, for values not exceeding a few million years.

If, to the constants appearing in formula (4), we substitute their values, we obtain

$$T = \frac{1,52 \cdot 10^{10}}{\sqrt{t}} \quad \text{degrees} \cdot \tag{5}$$

from which it can be seen that when t is, for example, equal to one second the temperature is, as could be imagined, frighteningly high: of the order of $10^{10}$ degrees; but it decreases very rapidly so that, after one thousand seconds, it is already reduced to having the order of magnitude of one hundred million degrees already sufficiently low for intense nuclear phenomena to begin to cease taking place.

The temperature, therefore, varies in time with a well-determined law and with an equally determined law the pseudo-radius of the Universe varies (it is proportional to the square of time) and, therefore, only one parameter remains indeterminate: the density of neutrons. By playing on this one arbitrary parameter Gamow proposed to succeed in predicting the distribution of the abundances of the elements and in fact in a certain sense he succeeded. He succeeded, really, as long as one limits oneself to a very crude analysis of his results, but as soon as one tries to go into the details, one immediately encounters difficulties, and one would probably encounter many more difficulties if one succeeded in carrying out this analysis, which is extremely complicated to do.

The first difficulties are already encountered in the lower part of the periodic system as soon as one wonders a little in detail how the elements are gradually being formed. As already mentioned the first nucleus to form will be that of hydrogen then by aggregation of a proton with a neutron deuterium will form then by addition of another neutron hydrogen three, which will decay through a beta process into helium three. By addition of a new neutron helium three will transform into helium four. Already here a small difficulty arises because helium three by capturing neutrons tends rather to break up than to form helium four, however one can always think that at least a small fraction of helium three is transformed by capture of a neutron into helium four. But at this point the difficulty that arises is much greater because the nucleus of mass five does not exist. If one tried to form it by addition of a neutron to helium four it would immediately fall apart, creating an insurmountable barrier that prevents the subsequent formation of elements by neutron addition.

Actually, some way can be found to jump over, or rather to work around this barrier, and it is the following: in accordance with the formulas we wrote earlier, in the period of time in which these phenomena are supposed to take place, the temperature, although it has already greatly decreased, is still of the order of $10^8$–$10^9$ degrees, and at such high temperatures thermonuclear reactions can still take place, in a conspicuous way, reactions, that is to say, produced by the collisions between nuclei moving under the effect of thermal agitation. Thus it is not impossible to think that a nucleus of mass six could be formed, without a nucleus of mass five already existing, by a deuterium nucleus reacting directly with a helium four nucleus. In reality this reaction is extremely improbable, but not impossible, so that it is not excluded that a small amount of lithium six will be formed allowing, by subsequent neutron additions, the formation of heavier nuclei. And many more difficulties of this kind are encountered, for example, even nucleus eight does not exist in any stable form, and this missing step will have to be overcome by an artifice of the kind already described.

But the difficulties do not end there. Another one that cannot go unmentioned is that, if one postulates an initial neutron density large enough for them to form heavy elements in not infinitesimal quantities, one finds that the proportion between helium and hydrogen has nothing to do with the real one: that is to say, one would have more helium than hydrogen in contrast to the experimental evidence.

It only remains, then, to sadly conclude that this theory is incapable of explaining the way the elements came to be formed, which was after all to be expected.

It is proper, however, to give credit to Gamow for the courage with which he set out to construct an attempt at a theory based on extremely determinate assumptions. The theory has failed which means that some of the hypotheses are wrong. But the result achieved in this way (of being at least certain of having made a mistake) is certainly more remarkable than what could have been achieved through a theory so vague as to be capable of explaining a considerable number of experimental facts, precisely through the large number of arbitrarinesses in it, but not of highlighting which of its points are certainly wrong, allowing them to be corrected and new and more satisfactory theories to be constructed.

## Biographical Notices on Ettore Pancini

### AT and PT

Ettore Pancini was born in Stanghella (now in the Province of Rovigo) in 1915. He attended elementary and lower secondary school in Venice. His secondary education was completed at the Military College in Naples. He was a student of Physics at the University of Padua where he graduated, at the end of 1938, with Bruno Rossi as his supervisor (1905–1993), replaced by Angelo Drigo (1907–1978) when Rossi was compelled to leave the University and Padua on October 1938, because of the Fascist racial laws.

At that time Pancini met Eugenio Curiel (1912–1945), who graduated in Physics at Padua with Bruno Rossi, a member of the clandestine Communist Party in Venice. Until 1956 Pancini was an authoritative member of the party which he left in 1969, to join the new leftist group of Il Manifesto magazine. After Rossi's forced departure, Pancini was named assistant professor at the physics Institute of the University of Rome on December 1940.

He joined the Cosmic Ray group under Gilberto Bernardini (1906–1995), professor of advanced physics at the University of Bologna, who worked in Rome a few days a week. In that period Pancini, in collaboration with Gilberto Bernardini (1906–1995), Marcello Conversi (1917–1988), Eolo Scrocco (1916–2012) and Giancarlo Wick (1909–1992), carried out measurements dealing with the surplus of positive mesotrons ($\mu$ meson) with respect to negative ones. Again with Bernardini, Conversi, Scrocco and Wick he measured the anomalous absorption of mesotrons in low density media compared to high density media.

Pancini's researches were interrupted in February 1941 when he was called up to the Army as a second lieutenant of anti-aircraft artillery. In the winter of 1942–43 he obtained permission that allowed him to work for a few months at the Plateau Rosa with Gilberto Bernardini and Oreste Piccioni (1915–2002). In September he was active in the Veneto as commander of the Partisan Action Groups of Venice.

After the liberation of the North of Italy on 25 April 1945 Pancini came back to Rome where Bernardini was working on the idea of building a high altitude laboratory for the study of cosmic rays. Pancini became Bernardini's closest collaborator and was named director of the Testa Grigia laboratory, a position held until 1953.

In 1946 Pancini returned to his work with Conversi and Piccioni who, in very difficult conditions, had developed a sophisticated technique for use in direct measurement of the mean life of the mesotron. Pancini came back to Rome just in time to participate in the collection of the data of the experiment which proved that the mesotron was not the Yukawa particle, now known as the Conversi Pancini Piccioni experiment, often quoted by Fermi in his 1949 Fermi lectures given in this book. The particle postulated by Yukawa was discovered in 1947 in Bristol by Cecil F. Powell (1903–1969) and Giuseppe Occhialini (1907–1993): the pion $\pi$, which decayed into the mesotron.

In 1950 Pancini obtained the chair at the University of Sassari. In 1952 he went to the University of Genoa, to the chair of experimental physics of Giuseppe Occhialini, who had moved to the University of Milan, the chair previously held by Augusto Occhialini (1878–1951).

In 1961 Pancini moved to the University of Naples where, besides his commitment to physics teaching and research, he took part in a research group on the teaching of science at the elementary level. Moreover he was involved in the International Genetics and Biophysics Laboratory, which had started its activities in 1962 under Adriano Buzzati Traverso (1913–1983).

In 1979 Pancini entered a research group which built instrumentation in Naples transferred to Adone in Frascati.

He died in 1981.

For further information see: Alvarez (1987); Amaldi (1989); Conversi et al. (1947); Gariboldi (2014); La Rana (2015); Liquori (1981).

## References

Alvarez, Luis W. (1987). Adventures of a Physicist (New York: Basic Books, Inc., Publishers, 1987).

Amaldi, Edoardo (1989). "Ricordo di Ettore Pancini' Il Nuovo Saggiatore 1989 5: 34-46. Pubblicato anche in Battimelli, Giovanni; Paoloni, Giovanni (eds) (1998). 20th century physics: essays and recollections. A selection of historical writings by Edoardo Amaldi (Singapore-New Jersey-London-Hong Kong: World Scientific, 1998), pp. 388-400 and pp. 401-414 (in English).

Conversi, Marcello; Pancini, Ettore and Piccioni, Oreste (1947). "On the Disintegration of Negative Mesons' Physical Review 1947 71 (3): 209-210.

Conversi, Marcello (1981). "The period which preceded and led to the 1946 discovery of the leptonic nature of the "mesotron": some personal recollections' Preprint CERN-EP/81-09 29th January, 1981. http://cds.cern.ch/record/127361/files/198102196.pdf accessed in September 21, 2023.

Gariboldi, Leonardo (2014). "Pancini, Ettore' in Dizionario Biografico degli Italiani (Roma: Istituto della Enciclopedia italiana, 2014), vol. 80. https://www.treccani.it/enciclopedia/ettore-pancini_(Dizionario-Biografico) accessed in September 19, 2023.

La Rana, Adele (2015). "Ettore Pancini: un'avventura scientifica e umana' https://people.na.infn.it/~celebrazioni/A000%20-%20File%20originali/057%20-%20Pancini/05_La_Rana.pdf, accessed in September 9, 2023.

Liquori, Alfonso Maria (1981). "Ettore Pancini e la Scienza a Napoli negli anni sessanta' Le Scienze 1981 159: 5-8.

# Fourth Lecture The Search of an Attraction Between an Electron and a Neutron

**Enrico Fermi**

## edited by N.B. Cacciapuoti

A problem that arises every time you have two particles is to know which forces act between them. Now the elementary particles whose existence is established are nine—as we have said. This means that the number of such problems is

$$\frac{9 \cdot 8}{2} \text{ namely } 36.$$

It should be kept in mind, however, that these thirty-six problems cannot always be solved. In fact, the solution of the problem is only possible in a limited number of cases, and the classical method of finding which forces act between two particles—or at least the most classical method—is to study the collision between the particles themselves.

In fact, already in Rutherford's famous research on the action exerted by atomic nuclei on alpha particles, the application of this method made it possible to establish that the size of the atomic nucleus is small compared to the size of the atom. And this method is still being followed with considerable success, although it gives nothing but partial results, to determine the forces between neutron and proton or between proton and proton, from the empirical point of view.

But the problem I propose to discuss today is a problem of much less importance. It is also a problem concerning the forces acting between two elementary particles, however, the two elementary particles considered are such that very few and unimportant forces are exerted between them.

(*) Held on October 10th, 1949, in the lecture hall of the Physics Institute of the University of Rome.

E. Fermi (✉)
University of Chicago, Chicago, IL, USA

 
A. Treves and P. Tucci (eds.), *Enrico Fermi, Atomic Physics Lectures*, History of Physics,
https://doi.org/10.1007/978-3-031-68690-0_7

The two particles in question are the neutron and the electron. So the problem we propose to discuss is "the forces between a neutron and an electron".

Of course, there is no Coulomb force in this case because one of the two particles is neutral. There is, however, a force of electrical or electromagnetic origin, the existence of which is known and has been observed for about a decade, which is due to the following circumstance: the neutron is endowed with a small magnetic moment. Consequently, when it passes through a magnetic field it is subject to forces because there is a mutual energy between this magnetic moment and the magnetic field. Since the electron, by its movements, can produce a magnetic field, a magnetic action between neutron and electron ensues. This fact was first suggested theoretically and then observed experimentally by Bloch, who deduced some very important consequences, and in particular a method for polarizing neutrons, that is to say for obtaining a beam of neutrons all having their spin oriented in a certain direction.

On the other hand, for what I intend to discuss today, this is not the kind of action with which we shall be concerned. Indeed, the circumstances under which the experiments we shall consider take place will always be chosen so that such action is eliminated. This can be done by always choosing for these experiments non-magnetic materials—atoms that are not paramagnetic, much less ferromagnetic, and there is nothing with which the magnetic moment of the neutron can interact. The choice of these particular experimental conditions is made deliberately because our aim is to research whether forces of an attractive character exist between neutron and electron.

Consider a neutron N and an electron E.

For these two particles, about which not much is known, it may be that there are forces acting between one and the other. So that, apart from any theoretical assumptions that can be imagined, the following problem remains open: do forces of an attractive or repulsive character exist, forces having essentially the character of central forces, between these two particles?

There is, however, in this case, a more specific argument for suspecting the existence of forces of this kind and it is as follows: the neutron, in fact—according to Yukawa's theory—is not really and strictly speaking a neutral particle because—according to this theory—a continuous exchange takes place, with a continuous reaction in opposite directions, which we have already written about in the following manner:

$$\mathrm{N} \rightleftarrows \mathrm{P} + \pi^- \tag{1}$$

the neutron can transform in either direction as follows: into a proton plus a negative meson, which, in this case, would be a $\pi^-$ meson.

And Yukawa's idea is that a continuous exchange takes place between these two forms of the neutron (they are both neutral forms, of course), so that the particle spends a certain fraction of time in one form, and the complementary fraction in the other form. Now there are some indicative reasons to believe that the particle is 80%

of the time in the N form and about 20% of the time in the other form i.e. P+$\pi^-$.[1] These figures are purely indicative and will be used only for an evaluation of orders of magnitude.

If, therefore, we wish to represent the neutron with a very rudimentary model, we can say that sometimes the neutron presents itself as a particle with no charge and sometimes the neutron turns into a proton having in its vicinity a negative meson, in accordance with formula (1). Of course, this negative meson may be found anywhere with respect to the proton. It will be found on average with a distribution having spherical symmetry around the proton so that we can describe the neutron from a statistical point of view as a particle defined by the mean value of the events that occur.

That is to say we can distinguish it in the following way: in the center we have, on average, a positive charge, because during 20% of the total time we have a proton in the center: therefore we have in the center a charge, if I may so say, of 0.2 times the elementary charge. Around this central positive charge, at varying distances, and at any position, there is instead a negative meson, which gives a kind of negative electricity distribution, which has, in total, the same value, but this time with a minus sign: – 0.2 times the elementary charge *e*.

**Fig.1**

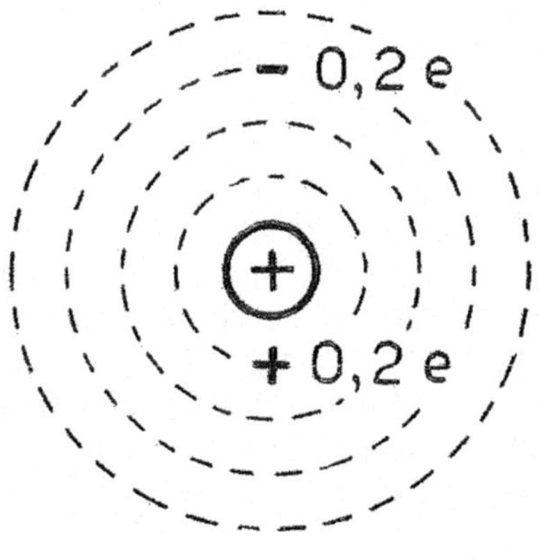


And this can be a rough electrical model of how this neutron should be understood. So the neutron is neutral in the sense that the total charge is zero. However, the electric field is not zero everywhere, because the positive charge is localized in the center, while the negative charge is spread spherically around this center (Fig. 1).

[1] One can get a rough indication of this fact from considerations of magnetic moments. The magnetic moment of the neutron is given approximately by $-2\ M_0$, with $M_0$ being the nuclear magneton. In contrast, the magnetic moment of the meson is of the order of $-10\ M_0$ (the mass of the meson is in fact of the order of 1/10 of the nucleon mass). According to the model we contemplate here it must therefore be admitted that the meson exists only 20% of the time, so as to give rise to an average magnetic moment equal to $-2\ M_0$.

So if we explore the electric field of this distribution of electricity, when we are far from the particle the electric field evidently cancels out and the potential is zero, but, when we approach a point that penetrates into the interior of this cloud of negative charge, then at this point the action of the central positive charge of value 0.2 $e$ is felt, not completely shielded by the negative distribution of charge so that at this point there is an electric field. And this electric field becomes stronger the nearer one gets to the center. Indeed, near the center, where the shielding action practically comes to an end, the electric force will be that due to a charge of 0.2 $e$.

If now the neutron is invested by an electron, according to this model, we can expect that, when the two particles come into extreme proximity to each other, forces are exerted between them. Actually, if this model is correct, the force must be attractive and of very short range because, of course, the mesonic cloud around the center is of very small radius. In a previous lecture I have already pointed out that the distances of the meson from the center are of the order of magnitude of the so-called Compton wavelength calculated for the mesonic mass (and not for the electron mass). Thus, the radius of distribution is of the order of magnitude

$$\frac{\hbar}{\mu c} = \frac{h}{2\pi\mu c}$$

where $h$ is Planck's constant and $\mu$ represents the mass of the meson. The radius is thus of the order of magnitude $10^{-13}$ cm.

Thus it follows that if this mechanism has actual physical significance, these are forces that reach up to a distance of only $10^{-13}$ cm. Indeed, one can at this point speculate on the following question: the electron itself probably cannot be regarded as a point. Even if we accept the classical idea of electro-magnetic mass, we must attribute to the electron linear dimensions that are of the order of $10^{-13}$ cm, indeed perhaps two or three times larger. We can therefore schematically represent the electron and neutron with two spheres whose radii have dimensions that have a ratio one to the other of about 2:1 or 3:1, as is shown in Fig. 2, although such a representation does not mean much.

**Fig.2**

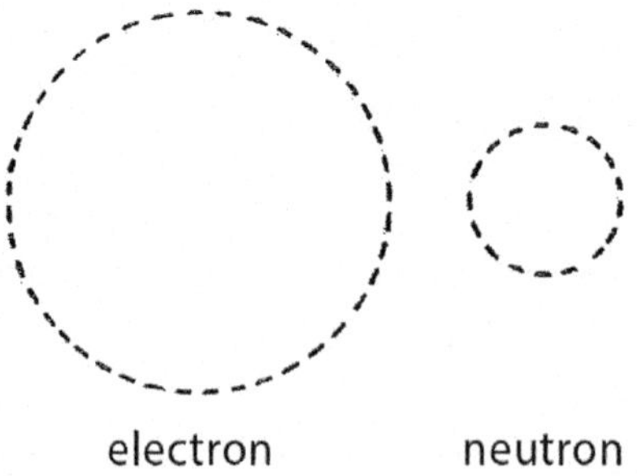


So if we take this representation somewhat literally, probably more than is permissible, we must say that already when certain parts of the electron begin to penetrate within the mesonic cloud surrounding the positive charge of the neutron, forces begin to act.

Now, the main problem, which is the object of the research I will expound, is to look for two things: first, whether by chance there are not forces acting between electron and neutron much greater than the one we have considered. Then, if this is not the case, to see whether we can observe the existence of the force whose origin we have now described.

It is clear that in order to understand the kind of experiment that can be carried out it pays to know the order of magnitude of the values we presumably need to arrive at, and this is fairly easy.

I must make an observation in this connection: in order to ascertain or not the existence of a force of the kind considered by means of collision experiments, it is necessary to know from theory what is the effect of a potential of the kind that intervenes in the model now described, what kind of collision phenomena it causes, and what kind of cross section it determines. Now this is a very classical quantum mechanical problem which, in the case of fairly weak forces of small magnitude, as in our model, gives the following result: the cross section depends not on the details of this potential which we shall denote by $u(r)$ but simply on the integral of $u(r)$ extended to the whole volume V. We find precisely that the cross section $\sigma$ is given by the following formula:

$$\sigma = \frac{\mathrm{m}^2 b^2}{\pi \hbar^4} \quad \text{with } b = \int_{\mathrm{V}} u(r)\, d\mathrm{V}. \tag{2}$$

In this formula $m$ is the reduced mass of the system, and we will see that this has some importance.

So in order to solve the problem of determining what the cross section would be in the collision between an electron and a neutron, according to the model we have described, we have to calculate or estimate the order of magnitude of the entity we have denoted by $b$. Such an estimate is not difficult (I am now making a calculation that is purely an estimate, forgetting all the numerical factors, which we will then put at the end as best we can), because the order of magnitude of the potential is given by the product of the central charge, which is 0.2 $e$ for the charge $e$ of the electron, divided by the distance between these two charges. To arrive at $b$ one must integrate this potential to the whole volume. The order of magnitude of this integral is given essentially by the potential multiplied by the volume. Apart from a numerical factor the volume is given by the cube of the radius of the charge distribution. Again, for an evaluation of orders of magnitude we can give both the radius of the distribution and the distance between the two charges considered, the value $\hbar/\mu c$ representing, as we have seen, the radius of the forces. In conclusion we find

$$b \approx -\frac{0,2 \cdot e \cdot e}{\hbar/\mu c}\left(\frac{\hbar}{\mu c}\right)^3 \quad \text{namely} \quad b \approx -0,2 \cdot e^2\left(\frac{\hbar}{\mu c}\right)^2 \tag{3}$$

One may ask: can one do better than this estimate? We shall see that at the present time the experiments are still extremely uncertain, so that at the present time one can only make a comparison of orders of magnitude. But the experiments will probably be refined, so that the problem will arise of comparing the result obtained from the experiments with that obtained from the theory more accurately than is required by a crude comparison of orders of magnitude.

At the present time, as I have said, one cannot do better than to be content with this comparison of orders of magnitude. And the reason why we cannot do better is, as usual, as I have pointed out on other occasions, that Yukawa's theory is a theory that has two flaws: first of all, it is not one theory but half a dozen theories, and we have to say more than half a dozen, because we do not know the seventh one and it is probably the good one. Secondly, even if we knew that one of these six theories that we do know is the good one, even then our mathematical methods of dealing with field theory (again apart from the indefiniteness of the theory) are extremely rudimentary and give rise to infinities that are sometimes accepted, sometimes eliminated rather artificially. So that in many cases the six theories (not that there are really six) give indefinite results. Moreover probably the good theory is not one of these six.

So at the present time not much more can be given than a crude evaluation of orders of magnitude of this kind, and this evaluation, substituting in (3) the numerical values, gives as the result:

$$b \approx -9 \cdot 10^{-46} \quad \mathrm{erg} \cdot \mathrm{cm}^3. \tag{4}$$

One can see a little more significantly what this number means in the following way. According to the well known convention we have a certain potential well in the neutron and what matters is the volume integral of this well. Let us fix a certain standard volume to which we will always refer and give the potential an average value such that the product of the potential for the standard volume is equal to the value of $b$ given by (4). What is usually taken as the standard volume is the classical volume of the electron (Fig. 3). If this is done, it is found that the average value that must be assumed corresponds to a potential of about 6000 eV.

**Fig.3**

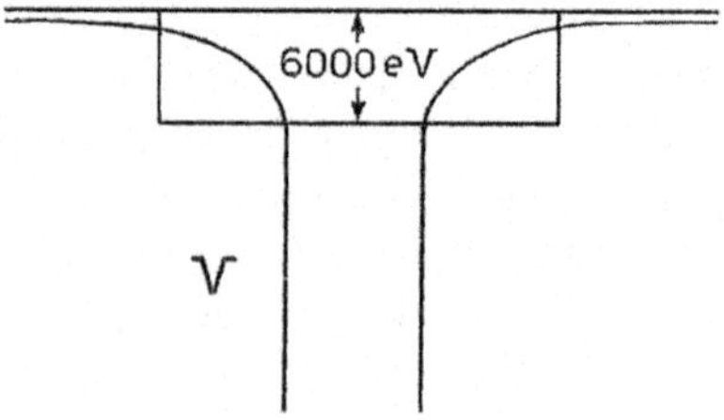


From this calculation we have an idea of how small this interaction is. We can compare it with the forces between nucleons, for example between protons and neutrons, which if standardized in the same way, that is to say conventionally interpreted as a potential well with a radius equal to the classical radius of the electron, correspond to potentials averaging around 15 million volts i.e., to values about 2000 times larger.

Thus, if such forces exist between electron and neutron they correspond in importance to about 1/2000 of the forces between nuclei.

But, for a moment, it may seem that the situation is even worse than that corresponding to this factor 2000. If we propose, for example (assuming that a force of this order of magnitude really exists), to find the cross section between electron and neutron we must use formula (2), which we transcribe again:

$$\sigma = \frac{m^2 b^2}{\pi \hbar^4} \tag{2}$$

for the value of the mass we have to put the reduced mass of the electron-neutron system, and since the neutron has much larger mass than the electron, the reduced mass is practically the mass of the electron, which is quite small and appears squared in (2).

Thus the result is small. To be more precise the numerical value is given roughly by $2 \times 10^{-37}$ cm$^2$ which is an extremely small cross section when compared with the cross section of neutrons with nuclei, which is generally of the order of $10^{-37}$. That is to say, there is a factor of $10^{13}$ of difference.

One wonders, then, how one can hope to observe an interaction corresponding to a cross section that is $10^{13}$ times smaller than the cross section of nuclei.

At this point we can begin to try to gain some factors so as to present the situation from a more plausible point of view. Meanwhile, there is a factor of one million, which can be put directly without difficulty, for this reason: in (2) $m$ appears, which is the reduced mass. But do we really have to take the reduced mass of the electron-neutron system? This would certainly be the case if the electron and neutron were free particles, that is if we only had an electron and a neutron colliding in a vacuum. But the electron is actually part of an atom to which it is attached virtually rigidly, with respect to the collision it receives when a slow neutron is deflected. Consequently (and this can be justified by careful calculation, but the result is understood well enough) an electron that is part of an atom is not a free electron and due to the effects of phenomena of this type behaves rather like a fixed electron.

So the system is now reduced to the collision of an electron, which, however, is pinned in a fixed position, against a neutron. So for the reduced mass that appears in formula (2) we have to take the mass of the neutron, which is about 2000 times larger than that of the electron, and it appears squared in that formula.

To our advantage this gives a factor of $4\times 10^6$ which is not enough, of course, to compensate for a factor of $10^{13}$ but it helps. We can also, without difficulty, gain another factor, perhaps 50, if we take a fairly heavy atom, having, for example, 50 electrons (various experiments, have been made with Xenon). These 2 factors together give a factor of $50 \cdot 4 \cdot 10^6 = 2 \cdot 10^8$ which begins to approach $10^{13}$. However, there still remains a factor of $10^5$ of difference and it is clear that one cannot do the experiments with the precision of one part in $10^5$. Experiments of the kind we are considering are quite difficult and it is already a great deal if we can get to one part in 1000.

But there is another factor that matters, and that is the factor of interference. For we must remember that phenomena of this kind must be treated with wave mechanics. Now from what was said earlier our problem is basically to study the collision of a neutron with an atom which has a nucleus and electrons. The wave representing the neutron is scattered both by the nucleus, giving rise to a certain wave, and by the various electrons, and for the moment we will consider one, giving rise to another wave. These two waves interfere, and since the action between neutron and nucleus is large while the action between neutron and electron is small, it follows that the amplitude of the resulting wave will consist of a large part A which is the amplitude of the wave originating from the nucleus, plus a small part a which will be the amplitude of the wave originating from the electrons. The intensity is proportional to the square of the amplitude of the wave. If there were only the nucleus the intensity would therefore be proportional to $A^2$, while taking into account the electrons it will be $(A+a)^2 = A^2 + 2Aa$ (neglecting a term $a^2$ which is very small).

On the other hand, the intensities are proportional to the cross section. So denoting the cross section of the nucleus alone by $\sigma_{nucl.}$ and the cross section of the electrons by $\sigma_{el.}$, we shall then have:

$$\frac{\sigma}{\sigma_{nucl.}} = \frac{A^2 + 2Aa}{A^2} = 1 + 2\frac{a}{A} = 1 + 2\frac{\sqrt{\sigma_{el.}}}{\sqrt{\sigma_{nucl.}}}$$

from which:

$$\sigma = \sigma_{nucl.}\left(1 + 2\frac{\sqrt{\sigma_{el.}}}{\sqrt{\sigma_{nucl.}}}\right) \tag{5}$$

Now we have seen that the ratio of the cross sections $\sigma_{el.}/\sigma_{nucl.}$ is $10^8/10^{13} = 10^{-5}$; however, in (5) fortunately the square root of this ratio multiplied by 2 appears, which gives as a result about 1/100.

We therefore see that due to interference, in which dispersion by the nucleus intervenes, the effects that can be expected are of the order of 1%, whereas we started from $10^{13}$, which would really have been out of the question.

Now, two experiments have been carried out to try to bring this phenomenon to light and have so far yielded only very partial results, purely by orders of magnitude, but they will probably be refined and eventually, I hope, a reliable result may be reached.

An early type of experiment was carried out by the Columbia University group headed by Rabi. Other experiments were made by L. Marshall and myself in Chicago. Both of these experiments are essentially based on the idea of observing the interference we have been talking about but the device is different in the two cases, and for brevity's sake, since I am more familiar with the experiment that I made, I will describe this one in more detail.

The idea of the experiment is as follows: consider neutron scattering by a noble gas. In the experiments performed, xenon was used as the noble gas.

Let us consider an atom consisting of its nucleus and peripheral electrons (Fig. 4) and try to observe the scattering of a beam of slow neutrons by this atom. The slow neutrons have a De Broglie wavelength, which is of the order of $2 \times 10^{-8}$ cm, i.e. of the order of magnitude of the linear dimensions of the xenon atom, which are $2\times 10^{-8}$, while it is several orders of magnitude larger than the size of the nucleus, which is practically point-like. The linear dimensions of the nucleus are about 20,000 times smaller than those of the atom and thus the wavelength of the slow neutrons.

Therefore, if there were no electron action (which we shall eliminate for the moment) and we were simply considering wave scattering by a relatively point-like nucleus, we would expect a distribution having exactly spherical symmetry, precisely because the wavelength is much greater than the obstacle constituted by the nucleus. But superimposed on this action of the nucleus, which is the main action, there must be the effect of the electron cloud, and those electrons are distributed, as I said before, over a region of the order of magnitude of the wavelength.

**Fig.4**

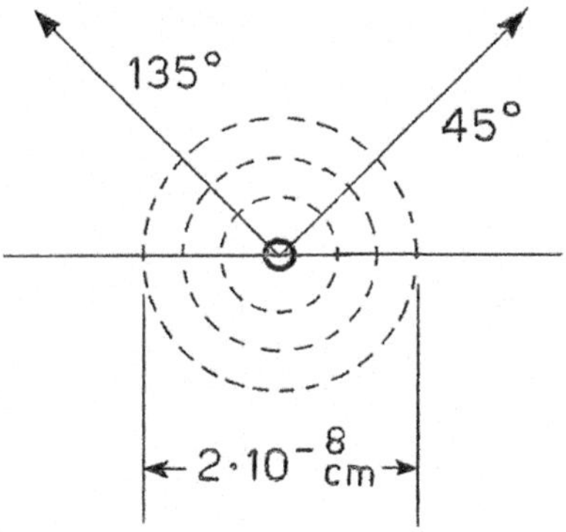


Therefore, if we observe the radiation of neutrons going in a certain direction, we will find diffraction phenomena, because the electrons are distributed over a region of the order of magnitude of the wavelength. Therefore, we can expect that, for example, by observing the intensity of neutrons scattered in a direction at 45° to the direction of the incident neutrons, or 135° to the direction of the incident neutrons, there will be differences. From the observation of these differences, we can trace what the effect of the electrons is.

To repeat: if there were no action with the electrons, the intensities in these two directions should be equal. The electrons will produce, because of the interference, a small difference.

The experiment was conducted as follows: it is one of those experiments in which it is very useful to have available the enormous neutron intensities that can be obtained from a uranium pile. The pile is surrounded by a large concrete screen. In this concrete screen there is a large cavity, a meter and a half in diameter, which penetrates into the interior of the pile. This cavity, however, is not empty, but is filled with graphite (Fig. 5). This graphite constitutes the so-called "thermal column". It is actually a column lying horizontally. In some machines, made at the beginning, this column was actually vertical, therefore it was called a "column" and this name was kept.

The purpose of this graphite is to purify the slow neutrons of the mass of neutrons of all energies that are produced in the pile. The neutrons enter the graphite where they begin to undergo a large number of collisions against the carbon atoms of the graphite and in these various collisions they lose energy. In this way, after diffusing through a relatively small layer of this graphite, all the fast neutrons are practically eliminated and at the exit of the thermal column there are only neutrons endowed with thermal agitation velocity, which are those with which the experiment can be made. In addition, it is necessary for the thermal neutron beam to be fairly well collimated, and for this we proceed in the following manner: a cylindrical hole is drilled in the center of the thermal column, along the axis, into which a cadmium tube is inserted. Cadmium is a substance that enjoys the property of strongly absorbing thermal neutrons: it absorbs them practically 100%. If we complete the outer end of the cadmium tube with a cadmium shield provided with a central hole, we obtain a collimator such that only neutrons passing through it can exit the hole.

**Fig.5**

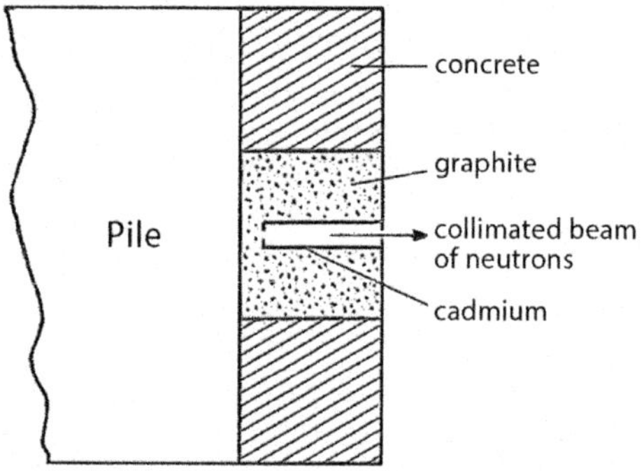


**Fig.6**

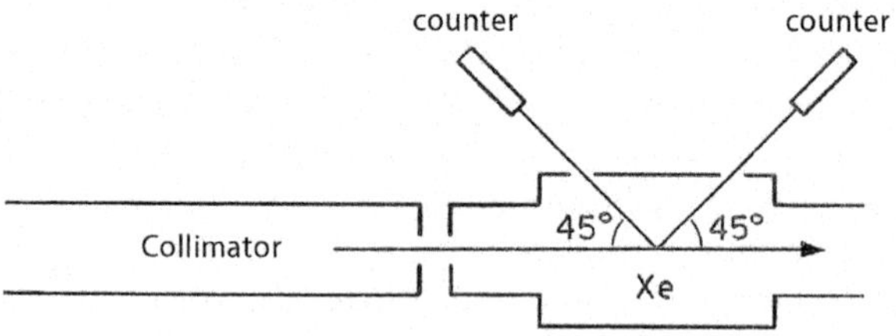


In this manner we obtain a collimated neutron beam of considerable intensity.

In Fig. 6 the neutron collimator is drawn again, together with the apparatus that is used for the measurements. The neutron beam is made to enter the actual apparatus, which is a kind of box that you can either empty or fill with xenon as appropriate. In the drawing in Fig. 6 only the part made of cadmium is drawn, which is the interesting part for neutrons. The whole thing is, of course, enclosed in a brass box in order to contain the gas.

In the upper part of the cadmium armor there are two windows beyond which two counters are placed which allow the detection of neutrons diffused at 45° and at 135° to the direction of incidence. Then the neutrons passing through the gas that is in this vessel undergo collisions so that some of them go into the one counter, and some of them into the other. Essentially by measuring the ratio of the number of neutrons arriving in the one or other counter we find the ratio between the intensity of scattering in the two directions.

There are a considerable number of corrections to be made, which I will not describe. However, if the full experiment is carried out, the result found is as follows: the ratio of the scattered intensity at 45° to the direction of the primary neutron to the scattered intensity at 135° turns out to be slightly different from one, and precisely this value is found:

$$\frac{I_{45^\circ}}{I_{135^\circ}} = 1,0235 \pm 0,0085. \tag{6}$$

However, this difference is largely due (perhaps for the whole, but certainly for the most part) to a spurious phenomenon. To be precise the following fact arises: xenon has an atomic weight of about 130, so it is much heavier than the neutron, which has an atomic weight 1. Nevertheless, the ratio of the masses is not infinite. If this ratio of the two masses were infinite, one could consider the collision as a process in which the xenon atom is stationary and the neutron bounces off it.

Actually, in this collision the center of gravity of the system moves slightly. One must then, with a very elementary calculation, limit ourselves to the case of a collision referred to the system of the center of gravity of the two particles. And if you do this you will find that the ratio should be 1.022, which is very close to the observed ratio.

So if we remove 1.022 from the experimental value of the measurement and if what is left is due to the effect we are looking for, and not to other small corrections, it is certainly a very small effect, where the probability of having forgotten something is not negligible. In any case, taking this result as correct, we can derive, let us say, the value of the potential well that corresponds to a depth of practically: $0 \pm 5000$ eV.

Thus, such an experiment, for now, gives a rather negative result because the value we expected was 6000 eV, albeit with considerable uncertainty. However, it should be borne in mind first of all that the theoretical estimate can easily be in error. Moreover, experimental errors are estimated on the basis of elements that are not always certain, so if we say that the error is 5000 in reality it may also be $10^4$, if there has been a series of accidents, all producing small errors in the same direction. It therefore seems clear that there is certainly no contradiction between the theoretical and experimental results.

As a future program I think this experiment can certainly be refined a great deal. In particular, I am convinced that the experimental error can be reduced so that we can get out of this unpleasant situation and find whether there is a principle of agreement between theory and experiment or not. At present what can be said on the basis of this experiment is that there is certainly nothing else very important, and the effect we are seeking may be there, or it may not be there. Future experiments will allow us to judge.

Rabi's method is based on essentially the same kind of theoretical idea, but the device is entirely different. In Rabi's experiment monochromatic neutron beams are used. In monochromatization of the neutron there is some loss of intensity so that the intensity is not sufficient to make an accurate diffusion experiment. Therefore, one is simply limited to a measurement of the total cross section.

What's more, again for intensity reasons, a gas cannot be used, so liquids are used (the experiments were done with bismuth and liquid lead). The experiment is to measure very accurately how the cross section varies as a function of neutron energy.

Now if one were to measure the effect due to a single nucleus with no electrons and no other nuclei, one should expect, in the energy region of about a hundred electron-volts, to find that the cross section remains constant as the energy varies. If there is a small effect due to electrons, the following fact must be taken into account: the wavelength of electrons is inversely proportional to their velocity, so it is inversely proportional to the square root of the energy. When one goes to high enough energies (thus to small enough wavelengths), the effect of the electrons is destroyed interferentially, because the wave phases associated with the various electrons surrounding the nucleus come on average to oppose each other. Conversely, when one goes toward lower energies, thus to progressively larger wavelengths, this interferential distribution of the electron effect gradually goes down: and this phenomenon gives rise to an effect that can be calculated. So it can be said that in the event that there is no electron effect we would obtain, in our approximation, that the cross section is constant as the energy varies. But in the event that there is this effect we should observe a small deviation of the cross section from this constant value.

There is an unpleasant point in this experiment. Actually, even the nuclei alone would not give a constant cross section, and this is because of the interference phenomena between the nuclei themselves. We are dealing here with a phenomenon very similar to what happens when x-rays are passed through a liquid. A liquid has properties of correlation between the positions of its various centers somewhat resembling those of a crystal; this gives rise to interferential phenomena. For this reason even the nuclei alone must give rise, and in fact do give rise, to a tendency in cross section as a function of energy that is almost constant for low energies but variable for higher energies.

It is clear, therefore, that the experiment to see the electron effect, if any, will have to be made in that energy zone where the perturbation due to the interference effect of the nuclei is very small. However, a residual of this perturbation is always an element of uncertainty in the experiment described, even if a very rough evaluation shows that this perturbation is in fact quite small.

Be that as it may, the impression is that Rabi's data result in a value of the potential that is probably negative and of the same order of magnitude as the theoretical value. More precisely, it seems that the experimental value may possibly be an order of magnitude lower, but certainly not higher than the theoretical value.

This, then, is the conclusion to which experiments of this kind lead. These are experiments in which, as I have said, very considerable improvements are possible so it is quite probable that in a year or two the situation will be quite clear. If this is the case, it will be possible to have greater familiarity with the mechanism we have described than we have today, according to which the mesons now and then leave, so to speak, the nucleus so that a neutron actually has a positive charge for a small fraction of time, while the meson is in its vicinity.

Qualitatively speaking, this result could make it possible to decide, at least in part, which is the true mesonic theory, or at least to exclude those theories that are found to contradict experimental facts.

Unless different and more precise methods of carrying out these experiments can be found, I do not personally believe that we can, with a simple refinement of current techniques, obtain much more than perhaps the $(-4000 \pm 1000)$ eV value.

This would already be an accomplishment probably close to the limits of what present experiments can allow but it would nevertheless be an accomplishment of some value, and work will be done in this direction to see how far we can go.

## Biographical Notices on Bernardo Ettore Cacciapuoti

### AT and PT

Bernardo Nestore Cacciapuoti was born in Toronto in 1913. He lived in Switzerland from 1920 to '27. Admitted as a student at the Scuola Normale Superiore of Pisa, he graduated in physics at the University of Pisa in 1936.

Then he became assistant to Emilio Segrè (1905–1989) at the University of Palermo, where Cacciapuoti and Segrè, with the advice of the chemist Carlo Perrier (1886–1948), identified three isotopes of element 43, named Tecnetium (Tc) in 1947. The three isotopes had been identified in a plate, sent to Segrè by letter, of molybdenum irradiated with deuterons in the University of California cyclotron. The chemical element with atomic number 43 was the first element produced in the laboratory by artificial means.

As a result of the promulgation of the 1938 racial laws, Segrè moved to the United States. Cacciapuoti also left Palermo when, in 1939, having been called as assistant professor to the chair of experimental physics of Edoardo Amaldi (1908–1989). In Rome Cacciapuoti concentrated his research on cosmic rays at sea level and on the distinction between electron and muon components. Between 1939 and 1944, despite the great difficulties of wartime, he lived perhaps his most scientifically happy period. At that time Cacciapuoti worked at the National Institute of Geophysics directed, between 1936 and 1949, by Antonino Lo Surdo (1880–1949). There Oreste

Piccioni (1915–2002) and Marcello Conversi (1917–1988), among others, worked and, together with Pancini (1915–1981), would make a decisive contribution to the knowledge of the $\mu$ meson.

In 1945 Cacciapuoti became the director of the newly established Institute of Physics of the University of Trieste. In 1962, he was called to the Chair of General Physics at the University of Pisa.

For his exceptional scientific skills and aptitude he was continuously called on to support diplomatic and scientific roles.

From 1951 to 1958 Cacciapuoti was vice-director of UNESCO's Department of Exact Sciences in Paris; he was scientific counsellor at the Italian Embassy in Washington from 1958 to 1961; he was a member of the Scientific and Technical Committee of EURATOM from 1962 to 1968; he was president of the Commission for International Relations of the Consiglio Nazionale delle Ricerche (CNR) (National Research Council); he was treasurer of the International Council of Scientific Unions from 1968 to 1974.

Although not Jewish, he agreed to teach experimental physics for the private Clandestine University of Rome, established in the academic years 1941–43 by Guido Castelnuovo with the task of preparing Jewish students, who were forbidden to enroll at State Universities, to take exams at the University of Fribourg. The program of his course, as well as that of the other courses, was the same as that held in the Faculty of Engineering of the University of Rome. The students of the Clandestine University were admitted to the third year of the Faculty of Engineering in January 1945, after the liberation of Rome from German occupation in 1944.

Bernardo Nestore Cacciapuoti died in 1979.

For further informations see: Agliolo Gallitto et al. (2020); Cacciapuoti and Segré (1937); Castelnuovo (2001), Foresta Martin and Calcara (2010); Gambaro (1987).

## References

Agliolo Gallitto, Aurelio; Chinnici, Ileana; Zingales, Roberto (2020). "1937: Palermo. The Discovery of technetium" in Esposito, Salvatore; Fregonese, Lucio; Mantovani, Roberto (2020) Atti del XXXVIII Convegno annuale SISFA - Messina 2018 (Pavia: Pavia University Press, 2020).

Cacciapuoti B.N., Segrè E. (1937)."Radioactive Isotopes of Element 43". Physical Review 1937 52: 1252-1253.

Castelnuovo, Emma (2001). "L'Universitá clandestina a Roma: anni 1941-'43 2 1942-'43"Bollettino Unione Matematica Italiana, Serie VIII 2001 IV-A: 63-77.

Foresta Martin, Franco; Calcara Geppi (2010). Per una storia della geofisica italiana. La nascita dell'Istituto Nazionale di Geofisica (1936) e la figura di Antonino Lo Surdo (Milano: Springer Verlag Italia, 2010).

Gambaro, Ivana L. (1987). "La scoperta del tecnezio"in Bevilacqua, Fabio (ed) (2020) Atti del VIII Convegno annuale SISFA - Napoli 1987 (Milano: Grafiche G.V., 2020), pp. 187-200.

# Fifth Lecture Nuclear Orbits

**Enrico Fermi**

## Edited by M. Ageno

One of the most important problems in nuclear physics is the following: knowing the number of protons and neutrons that constitute a given nucleus, calculate the binding energy of the latter, in its ground state. It is, of course, possible to broaden the program contained in this statement and propose to determine, in addition to the binding energy, all other quantities related to the nucleus in question, such as intrinsic angular momentum, magnetic moment, and so on.

An entirely analogous problem has already been encountered, and substantially solved, in the physics of the atom. In this case, given an atom, whose atomic number is known, we say that it consists of a central nucleus of given electric charge and a corresponding number of electrons. Here, too, one of the most important problems consists of determining the binding energy of this system in what is usually called its ground state.

The solution that has been given to this problem in the case of the atom is a partial solution, not so much because we do not know the principles on the basis of which to calculate an exact solution, but rather because the related mathematical calculation is too complicated to actually carry out. This partial solution is based, as is well known, on the following idea. The atom consists of the central nucleus of positive charge Ze and of Z electrons. Let us fix our attention on a chosen electron: the remaining (Z − 1) electrons constitute a kind of cloud of negative electricity surrounding the nucleus. The electron under consideration thus moves in the field created by the +*Ze* charge of the nucleus, and by the negative charge—(Z − 1) *e* of the remaining electrons. The approximation that is usually made, and which leads to some qualitative and often

(*) Held on October 12th, 1949 in the lecture hall of the Physics Institute of the University of Rome.

E. Fermi (✉)
University of Chicago, Chicago, IL, USA



A. Treves and P. Tucci (eds.), *Enrico Fermi, Atomic Physics Lectures*, History of Physics,
https://doi.org/10.1007/978-3-031-68690-0_8

also quantitative understanding of the properties of the atom, is to consider the field due to the electrons, which are many and rapidly moving, as a static central field.

In this way the problem is reduced for each electron to that of a single particle moving in an assigned potential field: this is a relatively easy problem to solve. Such an approximation is, of course, rather crude; however, there are two reasons why it is much better than it might seem at first glance. First, the contribution of the nucleus, which accounts for a little more than 50% of the field-generating charge, is described correctly in this approximation. What is incorrectly described is, of course, the contribution of the electrons, since it overlooks the fact that they are essentially moving point charges.

However, the forces that these electrons exert on the particular electron on which we have fixed our attention are Coulomb forces whose potential is $e^2/r$. These are therefore forces that decrease quite slowly as distance increases and thus are still noticeably effective at a relatively large distance from the particle from which they come. It can thus be said that a large number of electrons act quite intensely on a particular electron (at least in the case where Z is a large enough number), so that a statistical consideration of the kind indicated above, consisting of the substitution for the overall action of individual electrons that of a kind of cloud of static electricity, actually leads to a not too bad approximation.

One might now think of carrying out an approximate theory of the nucleus, which in particular would answer the question we asked at the beginning, starting with an idea essentially similar to this one. Consider a nucleus consisting of Z protons and N neutrons, and let us assume that both of these numbers are quite large. The forces exerted on a particular particle, say on a particular proton, come from all the remaining particles, that is, $Z - 1$ protons and N neutrons. One may ask whether the resultant force can also be represented here, as in the case of an atomic electron, by a kind of time-independent mean field.

Until very recently this question was always answered in the negative. Such a procedure, it was said, would be in the case of the nucleus far more crude than in the case of the atom, for the reason that the two arguments cited in its favor in the previous case are no longer applicable to the nucleus. On the one hand, one no longer has in this case a central charge that alone determines most of the potential energy. On the other hand, the forces exerted between nucleons are typical short-range forces. If, therefore, it were permissible for us to make a very crude model of the nucleus, consisting of a certain number of particles more or less packed in a certain configuration, it would follow that the particles acting on a particular one of the constituent nucleons are few, i.e. those that are in contact with it, if any sense can be given to this expression. The phenomenon is thus much less statistical in character than in the case of the atom.

This is what was being said until a couple of years ago or so. However, information of an experimental nature has gradually been accumulating that seems to indicate, in spite of the negative arguments now made, that even with such a schematization a certain number of results can be obtained. Of course, it is not yet known how far one can go with ideas of this kind.

Before delving into the subject, I would like to say a few words about a nuclear model that has been used, and is still used in many cases, and that allows us to arrive at a large number of approximate results. This is the drop model, which starts from the analogy between a nucleus and a drop of liquid, the molecules in the drop corresponding to nucleons.

**Fig.1**

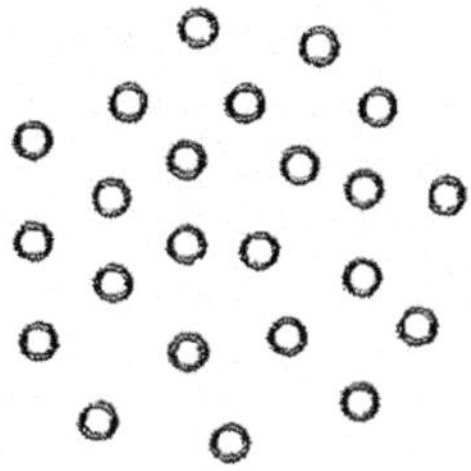

This drop model corresponds somewhat to a schematization such as that in Fig. 1. On the basis of it it will be possible to say that a particle that is in the interior of the nucleus is surrounded on all sides by nucleons. The binding energy will therefore contain a term proportional to the volume of the nucleus, that is to say, to the number of particles that constitute it. Particles, on the other hand, that are at the surface of the nucleus are subject to a different system of forces from the others, since they are in contact with only a fraction of the nucleons that usually surround one of these particles.

By developing these concepts, with very simple arguments of a semi-empirical nature, it is possible to write a very crude formula that solves quite well the problem we posed at the beginning, namely, to determine the binding energy of a nucleus. Instead of the binding energy, one can directly represent by means of this formula the mass of a nucleus, of mass number A and atomic number Z. The formula obtained is then as follows:

$$M(A, Z) = \underbrace{0,99391\,A}_{1} - \underbrace{0,00085\,Z}_{2} + \underbrace{0,014\,A^{2/3}}_{3} + \underbrace{0,083\,\frac{(A/2-Z)^2}{A}}_{4} + \underbrace{0,00063\,\frac{Z^2}{A^{1/3}}}_{5} + \underbrace{\delta}_{6}. \quad (1)$$

Term 1 of this formula, which is essentially proportional to the number of nucleons present, does not represent their total mass. In fact, the mass of a nucleon is equal to about 1,008 mass units. Instead, the coefficient of this term is smaller, due to the mass loss corresponding to the binding energy. Term 3 can be explained as an energy due to surface tension. It is in fact proportional to the surface area of the nucleus. As the radius of the nucleus is proportional to the total number of nucleons raised to 1/3, so its surface area is proportional to $A^2/3$. The coefficient 0.014 is determined empirically.

I shall not go into detail about the origin of the subsequent term 4. It evidently takes on a minimum value when $A = 2\,Z$, that is to say when the nucleus contains as many neutrons as protons. It thus has essentially the function of accounting for the well-known experimental fact that the most stable nuclear forms tend to be those containing equal numbers of these two types of particles. The term 5 has a very simple meaning: it is just the Coulomb energy. It is in fact proportional to the square of the charge and inversely proportional to the radius of the nucleus ($r \sim A^{1/3}$ ).

Finally, there is the term we denoted as δ and which is an erratic function of A and Z. More precisely, it takes into account the empirically known fact that the various nuclear forms are more or less stable depending on the equality or disparity of the number of protons and the number of neutrons. These numbers can be either both even or both odd, or one even and one odd, as is shown in the table below. Stability, other things being equal, increases markedly in the direction of the arrow.

| Z | N | | |
|---|---|---|---|
| $p$ | $p$ | | $\delta = -\frac{0{,}036}{A^{3/4}}$ |
| $p$ | $d$ | } ↑ | |
| $d$ | $p$ | | $\delta = 0$ |
| $d$ | $d$ | | $\delta = +\frac{0{,}036}{A^{3/4}}$ |

The values of δ that are chosen to represent the experimental facts correctly are given in the third column.

Formula (1), which is, as we have said, a semi-empirical formula suggested in its structure by the drop model, is actually quite simple, given the nature of the problem it solves; and it is also a very powerful formula, since it represents with surprising accuracy the nuclear masses through the whole periodic system.

However, one can, of course, propose to study the matter in more detail and ask how far this formula goes and what kind of errors are made by relying on it. It is found that the formula generally gives results in agreement with experiment. However, there remain irregularities of detail which cannot be represented by means of it. If, for example, we consider all the nuclei corresponding to a certain value of the Z/A ratio and represent the mass defects given for them by formula (1) in a graph, placing the mass number A in the abscissa, we obtain (apart from term 6 which we do not wish to deal with for the moment) a continuous trend of the kind depicted in Fig. 2.

**Fig.2**

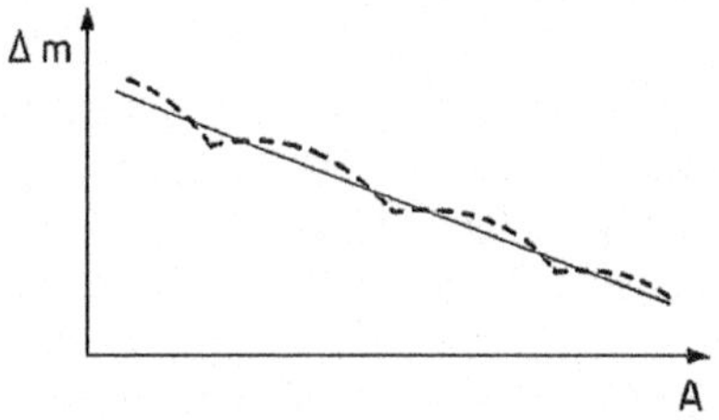


Now, it is found that the actual trend follows the calculated one quite well in general, although there are local deviations of the kind represented by means of the dashed curve in Fig. 2. The singular points correspond to well-determined values of the number of protons or the number of neutrons in the nucleus. These values, which are now commonly called magic numbers, are as follows: 2, 8, 20, 50, 82, 126.

I will not dwell too much on discussing the various properties of these magic numbers. I will merely point out, on the basis of Fig. 2, that they correspond to particularly stable nuclear forms. The experimental curve in Fig. 2, a section of which is shown enlarged in Fig. 3, is not actually a continuous curve. The abscissa represents a number of particles: thus the curve actually consists of a succession of points one unit apart on the abscissae.

**Fig.3**

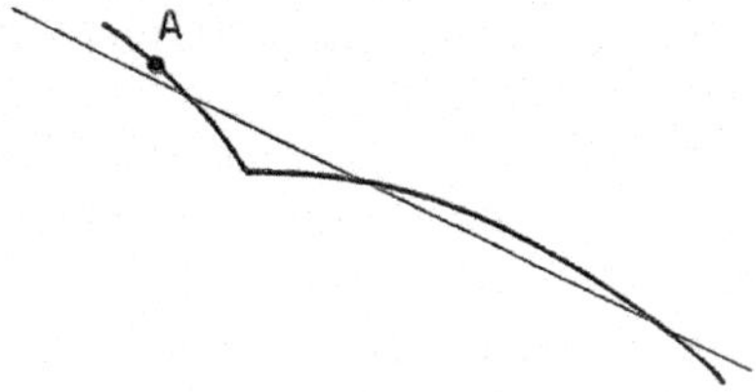


Let us start from a point such as point A, which is located just before a cusp. The mass defect for the addition of a particle in the nucleus increases strongly before the cusp: that is to say, up to this point, the individual particles in the nucleus are very strongly bound (Fig. 4).

**Fig.4**

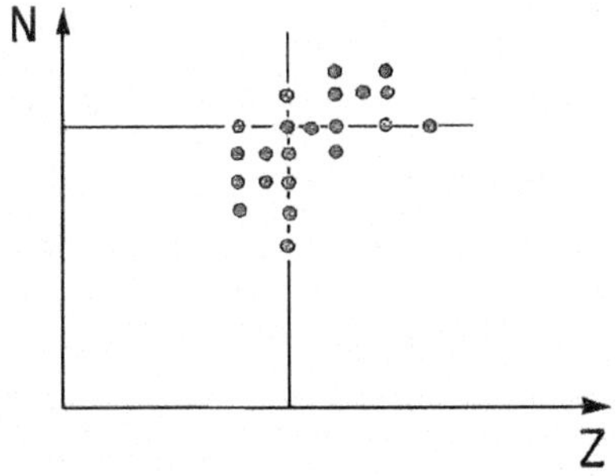


In contrast, this increase is much less immediately after the cusp. That is to say, after this point, the particles are at first much less bound, then as A increases they gradually become more and more bound until a situation similar to that at point A is again reached just before a new point of discontinuity.

In other words, you have a situation similar to that in the case of atoms. The K electrons here are the most bound, next come the L electrons, which complete the second orbit, always bound with some energy. The next electron is loosely bound but the binding energy increases as new electrons are added until another orbit is completed, and so on.

The experimental information on which the recognition of orbitals in the case of nuclei is based is largely of the kind given above. We represent stable nuclei in the usual diagram, placing the number of protons on the abscissae and the number of neutrons on the ordinates, and fix our attention on the number of isotopes corresponding to fixed values of N or Z. It is found that when, for example, the number of protons corresponds to one of the magic numbers the number of existing isotopes is much more numerous than usual. In fact, because of the strong stability of these nuclei, adjacent nuclei are transformed into them by beta transitions. The same happens when the number of neutrons matches one of the magic numbers.

One then also has data of another nature. If, for example, one examines the data concerning the cosmic abundance of the elements, one recognizes that nuclei in which Z or N is a magic number are particularly abundant. And one could, if one wished, cite another half-dozen facts of this nature. I will limit myself to mentioning only one, which refers to the behavior of very heavy nuclei and thus to the magic numbers 82 and 126. Seaborg, by bombarding with particles of very high energies produced with the giant Berkeley cyclotron, has recently greatly extended the alpha radioactivity table. In the extreme part of the periodic system containing the radioactive elements, he was able to produce and identify a huge number of nuclei emitting alpha particles and to determine the energy of these particles. And this not only for nuclei close to the stability curve, but also moving widely away from it on either side. Using Seaborg's data, by plotting the mass defects on the (N, Z) plane in a third dimension, it is possible by introducing very few assumptions to make a three-dimensional graph (a kind of map) of the trend of mass defects in this area, crossed by the magic lines corresponding to 82 protons, 82 neutrons and 126 neutrons. What is found is the three-dimensional analogue of the trend depicted in the two-dimensional case in Fig. 2.

Thus, if it is true that a situation similar to that of the periodic system of elements is also found in the case of the structure of nuclei, this suggests that in spite of the arguments to the contrary set forth at the beginning, one can also speak of orbits in a nucleus. The first attempt that comes to mind in this regard is as follows. The field acting on a nucleon inside a nucleus can be represented by means of what is usually called a potential well: the potential is zero outside an area that roughly represents the extent of the nucleus, while it rapidly falls to very low values inside. Along a diametral section of the nucleus the potential will then be represented by a curve with a trend similar to that in Fig. 5. One will then wonder what the sequence of orbits will be for a nucleon subjected to such a field. Of course, to give oneself more definite conditions it will be convenient to assume the potential well with sharp angles.

**Fig.5**

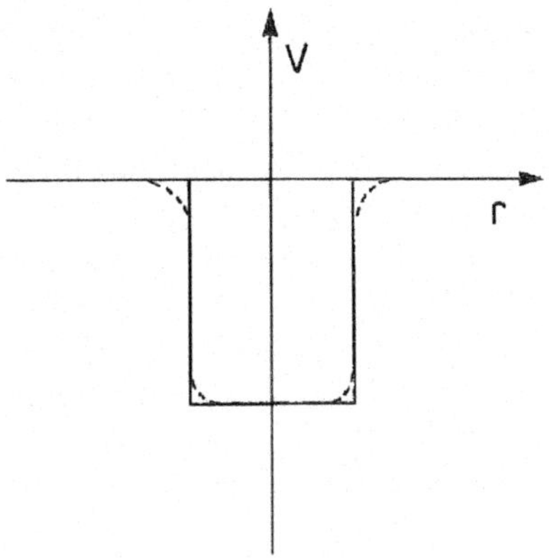


This problem is solved quite easily and leads to the following results. Here I use the spectroscopic notation to denote the areal moment of the orbits. By l I denote the areal momentum measured in h/2π units; then the letters *s*, *p*, *d*, *f*, *g*, *h*, *i* mean l = 0, 1, 2, 3, 4, 5, 6, respectively. The succession of orbits turns out to be as follows:

| orbit | number of particles |
|---|---|
| 1 *s* | 2 |
| 1 *p* | 6 |
| 1 *d* | 10 |
| 2 *s* | 2 |
| 1 *f* | 14 |
| 2 *p* | 6 |
| 1 *g* | 18 |
| 2 *d* | 20 |
| 3 *s* | 2 |
| 1 *h* | 22 |
| 2 *f* | 14 |
| 3 *p* | 6 |
| 1 *i* | 26 |
| 2 *g* | 18 |
| 3 *d* | 10 |
| 4 *s* | 2 |

If one does the calculation assuming a potential well with sharp angles, the various orbits, in this succession, are spaced at almost equal intervals. If, on the other hand, the corners of the potential well are rounded off a little, the intervals become unequal and orbit clusters are formed in the manner indicated by the parentheses in the above table. The second column of the table indicates the maximum number of particles that can be put into each orbit, according to the "Pauli principle". This number is calculated as in the physics of the atom and is given by $4l + 2$, $l$ being the azimuthal quantum number. For example, in *s* orbitals one can put two nucleons of the same type, neutrons or protons, in a *p* orbit one can put 6, and so on.

If the situation of nuclei were exactly as represented in the above table, the following predictions could be made based on it. The first orbit is filled with two nucleons, two protons or two neutrons—with that the first group is complete. So 2 is one of the magic numbers. The second group coincides with the second orbit and is filled with six particles, which added to the previous 2 gives 8 as the second magic number. The third group of orbits includes two orbits with a total of 12 particles: the third magic number thus turns out to be 20. So far as we can see things are going well and the predictions coincide with the experimental data. In the sequel, however, the model ceases to work, because the predicted magic numbers turn out to be 40, 70 and 112 instead of 50, 82 and 126.

In spite of this, R. Faenberg and Nordheim in two recent publications have tried to turn things around by admitting (actually quite arbitrarily) that the orbits in the above table can undergo appropriate shifts. Given our fairly meager knowledge of nuclear forces, there is evidently some latitude in the matter, but some of the reversals that they are forced to make seem entirely unreliable. Accordingly, I will not deal further with these attempts, and will certainly move on to discuss another one, proposed independently by Maria Mayer in Chicago and Suess. They start from the idea that there exists for the orbits of a nucleon in the nucleus a coupling between intrinsic moment and orbital moment, as for the orbits of an atomic electron. The spin of the particle can be oriented parallel or antiparallel to the orbital moment, so that actually two orbits come to correspond for every value of l, one with a resultant moment $l + 1/2$ and the other with a resultant moment $l - 1/2$.

It can now be admitted that a fairly large separation exists between these two energy levels. More precisely, it must be admitted, in order to be in agreement with the experimental results, that this separation is a fairly rapidly increasing function of l so that for small values of l it does not greatly alter the pattern previously described. Differences begin to occur for higher values of l and appear as follows.

The first three groups remain unchanged. The fourth group is joined by one of the two orbits into which the 1 g orbit of the fifth group is split, to be precise the one that has resultant momentum $4 + 1/2 = 9/2$. Similarly, one of the two orbits into which the 1 h orbit of the sixth group splits is added to the orbits of the fifth group, etc. In conclusion, the groups are modified as shown in the first column of the following table.

| Group | Orbits | Number of particles |
|---|---|---|
| I | I$s$ | 2 |
| II | I$p$ | 6 |
| III | I$d$ 2$s$ | 12 |
| IV | I$f$ 2$p$ I$g_{9/2}$ | 30 |
| V | I$g_{7/2}$ 2$d$ 3$s$ I$h_{11/2}$ | 32 |
| VI | I$h_{9/2}$ 2$f$ 3$p$ I$i_{13/2}$ | 44 |

Calculating as before the number of particles in each group, we now obtain the numbers in the last column. The magic numbers thus now turn out to be: 2, 8, 20, 50, 82, 126, as shown by experiment.

With this fairly simple hypothesis, one can then accurately predict the value of the six magic numbers. It could, of course, still be just a chance, and one can try to get more precise indications that the situation described is the actual one.

One actually has in this regard a number of quite indicative facts. I shall only mention the following. It is an experimental fact that all nuclei containing an even number of both protons and neutrons have zero intrinsic momentum. This means that if, for example, a neutron is added to a nucleus that has an odd number of neutrons the new particle will appear with the preexisting odd numbered neutron in the nucleus so that the resultant of their intrinsic momentum is zero. Let us then consider a nucleus for which either the number of protons or the number of neutrons is odd. For example, a nucleus with odd Z and even N. We can say that it consists of an even number of protons and an even number of neutrons, with the addition of an odd proton. It is reasonable to assume that all the particles, except the latter, associate with each other in pairs, so as to give a resultant angular momentum equal to zero. The angular momentum and magnetic moment of the nucleus will then be due substantially to the action of the odd proton. The same will be said if the odd-numbered particle is instead a neutron.

Consider then, for example, the nuclei of the kind mentioned above that are obtained by successively filling the orbits of the fourth group, which then have a number of protons (or neutrons) equal to any odd number between 21 and 50 and an even number of neutrons (or protons). If it is true that the angular momentum of these nuclei is due only to the odd-numbered nucleon, one can have for them only certain values of the nuclear spin and precisely those that correspond to the values of the resultant angular momentum of the various orbits. The following values will therefore be possible for the spin: 9/2, corresponding to the 1 $g_{9/2}$ orbit, 3/2 and 1/2, corresponding to the *p* orbits, 7/2 and 5/2 corresponding to the *f* orbits. These are thus the only spin values possible for the nuclei under consideration. The possibilities are indeed several, but the condition can be made much more restrictive by considering also the values of the magnetic moment.

The same value of nuclear spin can, for example, sometimes be obtained by adding the intrinsic moment of the odd particle ($l + 1/2$) to the orbital moment, or by subtracting this intrinsic moment from the orbital moment ($l - 1/2$). The value of the magnetic moment indicates, at least in many cases, whether it is the former or the latter circumstance, since this moment results differently in the two cases. If it is a proton, the azimuthal quantum $l$ is associated with a magnetic moment of $l$ nuclear units. If it is a neutron, on the other hand, the associated magnetic moment is zero, because the neutron has no charge and its orbital motion does not give rise to a magnetic moment. To this is then added in the case of a proton an intrinsic magnetic moment of about +2.7 nuclear units and, in the case of the neutron, a magnetic moment of −1.9 nuclear units. Combining the vectors with the Lande rule, one can then calculate what the magnetic moment of the nucleus should be in the various cases.

It cannot really be said that the experimental values agree too well with the theoretical values. If, for example, one plots the values of $l$ in abscissae and the values of the magnetic moment in ordinates, one obtains, according to this simple theory, two curves corresponding to the two different cases considered above. The experimental values, on the other hand, do not generally fall on the two curves, but are distributed in the plane approximately as shown in Fig. 6.

**Fig.6**

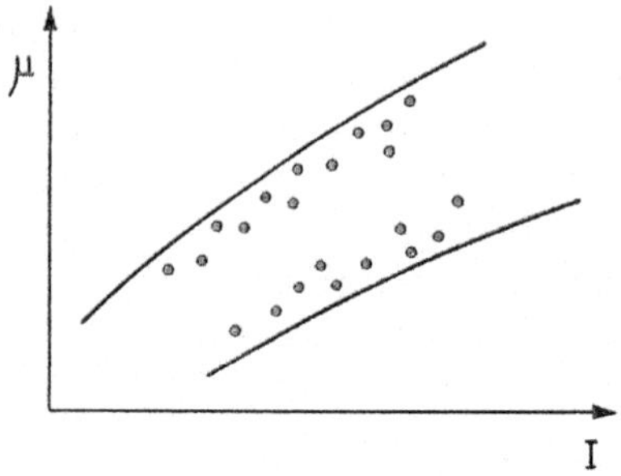


However, in many cases it is clear to which of the two categories the nucleus in question belongs, so that it can also be decided, in many cases without ambiguity, which $l$ values are acceptable.

Applying this procedure to the whole succession of nuclei, one finds about fifty cases where these data can actually be obtained, and in all these cases, with only a couple of exceptions, it seems that this description leads to a correct description of the angular moment and magnetic moment properties of the atomic nucleus.

So it seems that there is indeed something true in our model, although the model itself is certainly far from perfect. There is a very definite empirical indication that we can speak of nuclear orbits in the way I have indicated.

Thus several theoretical questions arise and first of all, since one can speak albeit approximately of nuclear orbits, one has to ask why the negative arguments mentioned at the beginning are not in fact as valid as we were inclined to assume. Moreover experimental facts indicate the existence of a strong coupling between spin and orbit responsible for that splitting of the levels that is necessary to make the magic numbers turn out right. One may wonder what the physical origin of this mutual energy is.

I can only mention here some still rather vague ideas that may perhaps contain some of the elements of the answer to these two questions. The fact that individual orbits for nucleons seem to exist in the nucleus could perhaps be justified in the following way. Consider (Fig. 7) an area of nuclear matter and a particular nucleon that is traveling along a certain trajectory. Evidently it will be legitimate or not to speak of this as a nuclear orbit, depending on whether one or the other of the following two facts occurs. If the average free path for a strong collision that abruptly alters the nucleon's trajectory is short, evidently the orbit will be completely destroyed. If this mean free path, on the other hand, is at least as long as the perimeter of the orbit, so that the nucleon can travel through this entirely at least once, then the orbit itself can be spoken of with some basis.

Fig.7

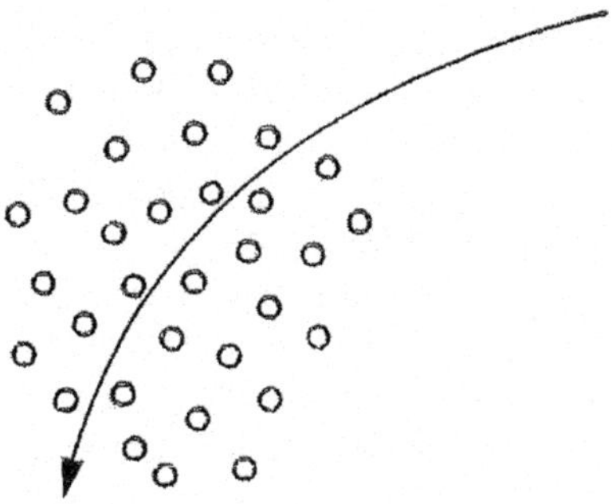

Fig.8

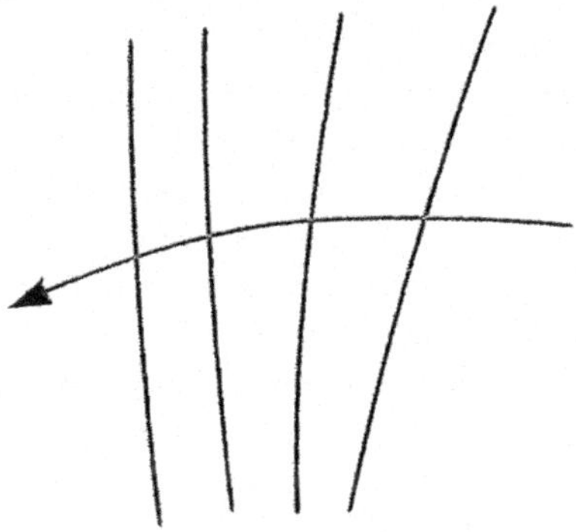

If one computes the average free path on the basis of cross sections between nucleons, one finds that it is indeed very short, which has many times been brought up as an argument that orbitals cannot exist. There is, however, one point that does not seem to have been considered in this evaluation, and that is the effect of the Pauli principle. This effect can be realized in the following way. The nucleon on which we have fixed our attention passes through a cloud of other nucleons, a cloud which at least in a certain approximation may be regarded as consisting of a degenerate gas of neutrons and a degenerate gas of protons. If a collision occurs it will be accompanied by an exchange of energy in which either the energy of the nucleon or the energy of the collided particle will decrease. But all this is forbidden by Pauli's principle, because either of these particles would fall into a zone of small energy whose levels are all occupied. It seems, therefore, that we must conclude that the mean free path should, indeed, be calculated on the basis of the cross sections of collisions, but discarding all those possibilities of collisions that contradict Pauli's principle. With this the number of collisions is greatly diminished and the mean free path is correspondingly lengthened.

This could therefore be, I say this with many reservations, the explanation of why nuclear orbits can, in spite of everything, be talked about. Then there is the question of the strong coupling between nucleon spin and orbit. This of course is a question of nuclear forces, which we can try to treat at least qualitatively on the basis of a mesonic theory. One such theory describes the mesonic field in a form that is quite similar to that of the electromagnetic field: it is the vector meson theory. If this were the correct theory, one could think of an interaction mechanism substantially similar to the coupling between intrinsic moment and orbital moment in the case of an electron. The latter is produced as follows: the electron moves in the electric field of the nucleus, whose lines of force are directed as shown in Fig. 8.

In the electron's reference system, everything therefore occurs, according to the well-known relativistic effect, as if there were a magnetic field equal to the electric field multiplied by the ratio of the nucleon's velocity to the speed of light. Since the electron has its own intrinsic magnetic moment, this is oriented and the mutual energy results differently depending on whether the intrinsic moment is parallel or antiparallel to this virtual field. Such a description can be translated into the language of vector meson theory by simply replacing the electromagnetic quantities with the corresponding mesonic quantities.

Unfortunately, an attempt to carry out a quantitative treatment of the subject, in this vector meson theory, cannot lead to anything more than an order of magnitude. Nor, of course, are there many reasons for believing that it really is the correct meson theory. However, I wished to mention this subject in order to show that there is indeed a possibility of arriving, tomorrow, by following the path indicated, at a consistent theory of all these phenomena.

Apart from these last theoretical considerations, the research we have been dealing with opens up a fairly large field of a semi-empirical nature. They make it possible in many cases to collect a good deal of data, for example, on the spin properties not only of the ground state of a nucleus, but also of its weakly excited states. If, for example, the spin of a nucleus is 9/2 and the orbit is presumably 1 $g_{9/2}$, the unexcited states of the nucleus are those in which the odd nucleon occupies one of the other orbits of the same group. The possibilities to be considered are thus rather small in number. Indeed, it may be noted that nuclei of this type often give rise to isomers. These occur when the fundamental orbit is, for example, 9/2 and the weakly excited ones are 1/2 or 3/2. Given the strong change in angular momentum, all transitions are then forbidden to a very high order, so that the weakly excited states have extremely long lives and thus give rise to isomers.

In conclusion, it can be said that this research opens up a chink of light to begin to construct a systematics of nuclei. Of course, before this theory can be considered in any way as definitive, the fundamental questions I have posed, namely, why nuclear orbits constitute a reasonably valid model and why there is a strong coupling between spin and orbit, will have to be fully clarified. The experimental material will also have to be examined with much greater care, in order to decide whether those regularities that are suggested by the model in question are actually, within certain limits, observed.

## Biographical Notices on Mario Ageno

### AT and PT

Mario Ageno was born in Livorno in 1915. He attended the first two years of the Physics Curriculum at the university of Genoa, and in 1934 he moved to the Physics Department of Rome University, at the time still located in the legendary via Panisperna. There Fermi and his colleagues had just completed their first experiments on neutron slowdown.

He obtained his Physics degree in 1936 with a thesis with Fermi as supervisor. In Rome he became familiar with Emilio Segrè (1905–1989) who, as Ageno recalled in 1989, published by Bonolis (2008), once said to the students that there they did not farm cabbages. If a student thought he was a smart person fine, he could stay. If not, it would be better to leave immediately.

Once in Palermo, Segrè asked Ageno to join him; but Ageno preferred to remain in Rome, where in the meantime he had been offered a position of assistant professor.

He started working on neutron physics having as mentors Edoardo Amaldi (1908–1989) and E. Segrè. For five years he was involved in the war. Coming back from the war for serious health problems, which afflicted him for many more years, he resumed his activity at the University of Rome in 1945 and participated in Amaldi's effort to restart physics research in Rome after the disruptions of the war years. He collaborated in the study of cosmic rays at the high altitude site of Testa Grigia.

In 1949 he moved to the Istituto Superiore di Sanitá (ISS) (National Institute of Health), where a Cockcroft Walton accelerator, originally designed by Enrico Fermi, E. Amaldi and Franco Rasetti was located.

Until 1969, when he left the Institute, most of his energies were devoted to the restructuring of the Laboratories, providing them with considerable means and personnel for the time. In the Istituto he organized and managed a research group which dealt with various issues: scintillation in liquids, quasi-elastic diffusion of electrons on a nucleus, synchrotron light, development of electron microscopy and radiation metrology, design and manufacture of original electron chains.

In 1969 Ageno left the Istituto Superiore di Sanitá for the University of Rome where he devoted himself fully to Biophysics, a borderland between physics and biology, maintaining, however, the attitude of a physicist rather than a biologist.

In 1973 he was appointed to the first chair of Biophysics in Italy. Biophysics was an old interest of his that had manifested itself since 1946 when he translated into Italian Erwin Schrödinger's *What is Life*? Ageno proposed a tool to sound out theories about the origin of life and differentiate those endowed with scientific plausibility from those that are not different from a creationist approach to the problem.

Speaking of Fermi's lectures, of which Ageno was the editor of the fifth Lecture, he later highlighted how those lessons, under the veil of extraordinary clarity, often concealed conceptual difficulties.

He died in Rome in 1992.

For further informations see: Ageno (1949, 1989); Amaldi and Maiani (1989); Bergia (2000); Cardinale (2017); Corbellini (1993–’96); Murzi (2010); Salvadori (1992); Schrödinger (1956).

## References

Ageno, Mario (1949). “Nucleo” II Appendice of the Enciclopedia Italiana. 1949.

Ageno, Mario (1989). “L’uomo Segrè: ricordi di un allievo” in Bonolis, Luisa (2008) Maestri e allievi nella fisica italiana del Novecento (Pavia: La Goliardica Pavese 2008), pp. 31-36.

Amaldi, Edoardo; Maiani, Luciano (eds) (1989). Fisica e biofisica oggi. Atti del Simposio in onore di Mario Ageno (Roma, 1-2 ottobre 1985) (Bologna: Societá Italiana di Fisica, 1989).

Bergia, Silvio (2000). “Mario Ageno, filosofo della fisica” Il Nuovo Saggiatore 2000 1-2: 23-28.

Cardinale, Adelfio Elio (2017). “Ageno e la nascita della biofisica in Italia” Nuova antologia 2017 618(2283): 209-223.

Corbellini, Gilberto (1993–’96). “Il dovere di essere scienziato” https://fisicamente.blog/2020/07/30/ricordo-di-mario-ageno/

Murzi, Mauro (2010). “Mario Ageno” in Pozzoni, Ivan (ed) (2010). Voci dal Novecento (Monza: Limina mentis, 2010) vol I

Salvadori, Paolo (1992). “Ricordo di Mario Ageno” Il Nuovo Saggiatore 1992 8(5/6): 30-42.

Schrödinger, E. (1956). What is life? (New York: Dobleday, 1956).

# Sixth Lecture New Developments in Quantum Electrodynamics

**Enrico Fermi**

## Edited by G. Morpurgo

Soon after the discovery of quantum mechanics the problem arose of extending it to the case of the electromagnetic field. It is a case where the extension is not obvious because the electromagnetic field, essentially, is a system with infinite degrees of freedom.

The first success in this endeavor was due to Dirac who first established a theory of radiation which accounted for the fundamental facts, already postulated in Bohr's old theory: namely, the relationship between frequency and energy levels, the intensity of spectral lines, etc.

Dirac's theory is essentially a direct generalization of the methods of quantum mechanics done as is well known according to the following scheme. If we consider the electromagnetic field in a cavity, in a box, to be precise, of finite size (let us denote its volume by $\Omega$), from considerations of statistics and electromagnetism it appears that this electromagnetic field is in many respects equivalent to an elastic medium filling the cavity and, as such, has certain characteristic frequencies. These characteristic frequencies have a discrete spectrum as long as the volume of the cavity is finite, whereas when that volume tends to infinity this discrete spectrum tends to a continuous spectrum. And indeed, said in passing, the reason why we imagine our field enclosed in a cavity is associated with the fact that it is easier from the analytical point of view to proceed with a discrete frequency spectrum than with a continuous spectrum.

(*) Held on October 14, 1949 in the lecture hall of the Physics Institute of the University of Rome.

E. Fermi (✉)
University of Chicago, Chicago, IL, USA

 
A. Treves and P. Tucci (eds.), *Enrico Fermi, Atomic Physics Lectures*, History of Physics,
https://doi.org/10.1007/978-3-031-68690-0_9

In particular, one result of the considerations mentioned above is the formula that gives what is the number dN of characteristic frequencies between $\nu$ and $\nu + d\nu$:

$$dN = \frac{8\pi}{c^3}\nu^2 d\nu \cdot \Omega \tag{1}$$

where $c$ is the speed of light.

That being said, Dirac's theory is based on the idea of considering each of these frequencies as the frequency of an oscillator of properties similar to those of a mechanical harmonic oscillator. The quantum mechanics of this system is then reduced to the quantum mechanics of an infinitely large number of harmonic oscillators, with the frequency distribution given by (1).

It will be good to note right now a difficulty that is not particularly serious, but which is perhaps a first indication of more serious difficulties to be encountered later on. It is known that according to quantum mechanics a harmonic oscillator, having a certain characteristic frequency $\nu$, has a succession of energy levels $E_N$, which are given by the formula

$$E_N = \hbar\nu\left(N + \frac{1}{2}\right) \tag{2}$$

where $N = 0, 1, 2, 3 \ldots.$ and $\hbar$ = (Planck's constant)/$2\pi$.

The formula (2) shows the existence of what is usually called the zero energy, that is to say it shows that even in the lowest energy state the oscillator has a finite energy precisely given by $\hbar\nu/2$.

If then we have not just one oscillator but several oscillators the overall zero energy will be the sum of the zero energies; and precisely if these oscillators are in a number of $8\pi/c^3\nu^2\Omega$ per unit interval of characteristic frequency, the total zero energy will be given by:

$$\frac{8\pi^2}{c^3}\hbar\Omega\int_0^\infty \nu^3 d\nu. \tag{3}$$

The integral in (3) diverges so that the zero energy of this system is an infinite energy. The one we have now mentioned is not the worst of the infinities that will be encountered later on, for the following simple reason: it is well known that in characterizing the energy of a system an indeterminate additive constant can always be left free. So in our case, to avoid infinite energy, it will be sufficient to redefine in a slightly different way the scale of energies, that is to say in the way that the energy of one of these oscillators instead of counting it from what is the conventional zero we count it from $E_N = 0$. The zero energy will thus disappear.

However, (3) is indicative of a type of mathematics that gives rise to most of those infinities which, as you certainly know, are one of the afflictions of electrodynamics and field theories; and the mechanism of such divergences is always roughly the same. It almost always involves doing sums extended to this infinite number of oscillators, and sometimes these sums (or integrals) give finite results while sometimes, and even often, they give infinite results.

We shall have occasion later to return to this point; and to continue now in the exposition of what was the development of quantum electrodynamics I shall say that, after this first attempt by Dirac to deal with the electromagnetic field, it was put on a fairly complete formal basis by introducing not only the radiation field corresponding to these oscillators, but also essentially the nonradiative part of the electromagnetic field, in a unique scheme, incidentally not very different from that now mentioned.

In this connection I would like to make an observation. Consider an electromagnetic field not already isolated, but with some charge present; and the simplest case will be one in which only an electron is present. The Hamiltonian of such a system can be written as the sum of three terms: the Hamiltonian $H_e$, of the electron as if the electromagnetic field were not there, plus the Hamiltonian $H_{em.}$ of the electromagnetic field as if the electron were not there, plus an interaction term $H_{int.}$, which of course will be proportional to the electron's charge and whose form results from the following formula, where $v$ is the electron's velocity and $\mathbf{A}$, $\varphi$ the vector and scalar potentials at the position occupied by the electron itself

$$H = H_e + H_{em.} + H_{int.} = H_e + H_{em.} - e\left(\frac{\mathbf{v}}{c} \times \mathbf{A} - \Phi\right); \tag{4}$$

$H_{int.}$ is effectively an interaction term in that electromagnetic field-dependent and electron-motion-dependent elements are simultaneously involved in it. Note that the interaction term has the electron charge $e$ as coefficient.

Now, in most treatments, the fundamental problem of the dynamics of our "electromagnetic field + electron" system which is to solve the Schrödinger equation whose Hamiltonian is given by (4) is treated mathematically by a process of approximation; and the approximation consists of considering $H_{em.}+ H_e$ as the unperturbed part of the problem and $H_{int.}$ as the perturbation. Thus we get the obvious advantage of having a very simple unperturbed problem because the electromagnetic field is not affected by the electron, and vice versa.

According to the normal prescriptions of perturbation theory all results are then obtained through a process of series expansion and the parameter with respect to which it is developed in series of powers in this case is the famous expression $e/\sqrt{\hbar c}$ Everything would be perfectly fine, or quite fine, if e were infinitesimal, but e is obviously not infinitesimal since it represents the charge of the electron. So substituting the numerical values we have

$$\frac{e}{\sqrt{\hbar c}} = \sqrt{\frac{e^2}{\hbar c}} = \frac{1}{\sqrt{137}}$$

and this expression is just under a tenth (which is quite small, or quite large, depending on taste).

I would like to add the following in this connection. Of course the quantum theory of the electromagnetic field was the first to develop because the electromagnetic field is certainly the field with which we are most familiar, but in the years between 1930 and 1940 there was a flowering of theories somewhat similar to this one in which other fields were involved. One such theory is the beta-ray theory which differs from this one because in that case the coupling constant is indeed extremely small (of the order of $10^{-10}$ or $10^{-13}$. In that case the series expansions may have some vague reliability. And another is Yukawa's theory of nuclear forces transmitted by the mesonic field in which, unfortunately, the expansion parameter is large, and actually about equal to one, or perhaps equal to one-half or perhaps equal to one-third.

In today's discussion, however, I shall confine myself to the case of the electromagnetic field, which has the great advantage over all others of containing no undefined element in what is the classical approximation, for the electrology is quite well established and certainly is known infinitely better than is the mesonic field theory.

Reconnecting with what I said above I would now like to come to some divergences, which are, as I have already mentioned, the focal point of all these theories. The problem is this: let us consider in the above-mentioned scheme an electron, and let forces due to the radiation field be the only forces acting on the electron (there are, in other words, no external fields present). It is well known, then, that in both quantum electrodynamics and classical electrodynamics the coupling between the electromagnetic field and the moving charge is presented as the equivalent of an electromagnetic mass, and indeed it is well known that in both the one and the other theories this electromagnetic mass turns out to be essentially infinite. The type of the infinity, however, turns out to be very different in the two cases. Essentially in the classical theory, the electromagnetic mass of an electron at rest that is not quite point-like but has a radius R, is given as an order of magnitude by the electrostatic energy and thus results from the well-known relationship:

$$mc^2 \cong \frac{e^2}{\mathrm{R}} \qquad \text{namely} \qquad m = \frac{e^2}{c^2\mathrm{R}};$$

therefore this electromagnetic mass becomes infinite as 1/R if we talk about a point electron.

In the quantum theory there is also an electromagnetic mass, the divergence of which, however, is of a different nature from the classical one. The reason for this divergence is related to Dirac's theory of the electron (existence of positive and negative energy states) and to the fact that a certain cancellation of terms occurs which makes the divergence very weak, as we shall see. To be precise, it is found that, if we consider an electron which has a mass $m$ of non electromagnetic origin, we can say a mechanical mass, it acquires, as a result of coupling with the field, an additional mass, which we shall call $\mathrm{m}_{el}$ to indicate that it is an electromagnetic mass. The mass we measure, the mass of the physical electron, $\mathrm{m}_f$, results from the sum of this electromagnetic mass plus the mechanical mass and the expression of the electromagnetic mass is as follows

$$m_{\rm el} \cong \frac{e^2}{\hbar c}\log\frac{\mathrm{E_{max}}}{mc^2}\,. \tag{5}$$

We wish to briefly discuss (5): $e^2/\hbar c$ is as usual the famous 1/137, then there is the mechanical mass as factor and finally the $\log(E_{max})/mc^2$. Concerning this logarithmic term we can say the following: it results from one of those usual sums extended to an infinite number of oscillators. In fact, if the sum were extended to all oscillators, the term itself would diverge. If we want it not to diverge we must make a "cut" i.e. extend the sum not to all oscillators but only to those whose energy is less than a certain $E_{max}$. Of course, such a procedure is in no way justified within the framework of a coherent theory. However, the fact that infinity is only logarithmic and that the energy $E_{max}$ of the cut can be chosen extremely large compared to the usual energies[1] $m_{el}$ authorizes one to hope that in a future theory (which possibly reduces to the present one for low energies) it will converge instead of diverging.

In any case in the present theory $\mathrm{m}_{el}$ diverges; and this difficulty is just a special case of the infinities that, in the studies of quantum electrodynamics that took place until a couple of years ago, appeared more or less everywhere whenever one wanted to be too ambitious.

The situation in which electrodynamics found itself a couple of years ago can be summarized essentially as follows: whenever one proceeds with the series expansion we mentioned earlier (corresponding to the perturbation method) to calculate a certain phenomenon the practical procedure was to stop as soon as a non-zero result was found because, if one dared to go on the following approximation was usually divergent. This is obviously not a very dignified way of conducting a calculation, but this was the rule of thumb that was advised to students.

Studies of this kind have been revived recently mainly as a result of an experiment by Lamb at Columbia University who has taken up recent techniques of microwave spectroscopic investigations of the hydrogen fine structure problem.

It is well known that if one considers, for example, the $n = 2$ term of the hydrogen spectrum, according to the theory one reads in books, for example in Sommerfeld's book, it consists of a term $2\mathrm{S}_{1/2}$ with total angular momentum 1/2; and of a pair $2\mathrm{P}_{1/2}$ and $2\mathrm{P}_{3/2}$.

The position of these terms according to what we read in Sommerfeld is as shown in Fig. 1 drawn here.

[1] The order of magnitude of $E_{max}$ can be determined in the following way: $E_{max}$ can be chosen such that it is of the order of $m$. If we do the math we see that it results in $E_{max} \simeq 10^{53}$ erg which is 100.000 times the mass of the earth translated into energy using the Einstein relation.

Fig.1

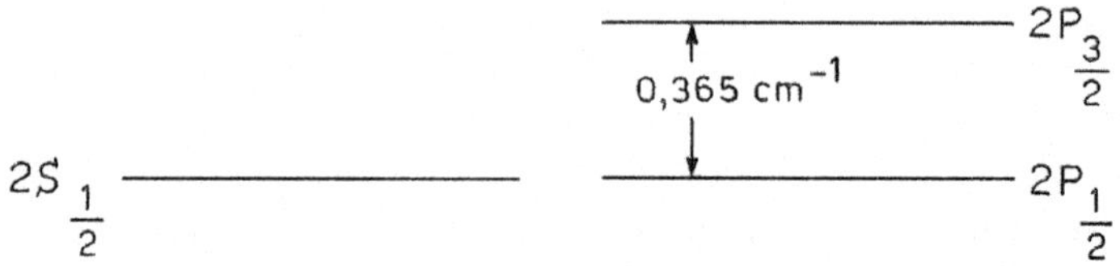


Fig.2

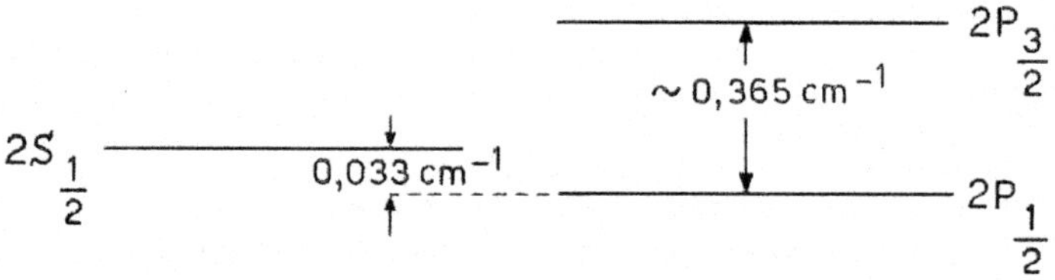


We have the two terms $2P_{1/2}$ and $2P_{3/2}$ separated by the frequency difference of $0.365\,\mathrm{cm}^{-1}$. Then we have a $2S_{1/2}$ term coinciding exactly with the $2P_{3/2}$. It must be said in fairness that the spectroscopists who had studied these phenomena had timidly expressed the opinion that this was not in fact the case; but since the experiments were not very precise, and in any case people probably had more faith in the theory than was legitimate, it had not attracted attention. Only when Lamb tackled the problem again together with the more powerful methods now available for measuring small frequency differences by means of microwaves, instead of with the systems of conventional spectroscopy, did everyone become persuaded that in fact the pattern is not as drawn now, and that, while the distance between the two levels $2P_{1/2}$ and $2P_{3/2}$ is to a great approximation that predicted by theory, the level $2S_{1/2}$ is shifted relative to $2P_{1/2}$ by something like $0.033\,\mathrm{cm}^{-1}$, which is about 10% of the total distance between $2P_{1/2}$ and $2P_{3/2}$ (see Fig. 2). It was precisely the attempt and the work that was done to explain this difference between theory and experiment that led to the truly remarkable progress that there has been in quantum electrodynamics in the last two or three years. It is difficult to specify to whom this progress is due because the origin of the work took place at a conference attended by about thirty theoretical physicists and in which the problem was discussed; and many of the general ideas that were later developed were, at least qualitatively, expressed without going so far as to develop calculations in detail. However, the first published work in which this situation was explained at least 99% was by Bethe, who calculated the effect of the perturbation of the $2S_{1/2}$ term due to coupling with all the radiation oscillators and by a trick of subtraction came to justify that the effect must be very close to, in fact almost exactly, the experimental effect.

In order to show Bethe's reasoning I should go back to (4) for a moment and put a few more things in place of the expressions He, Hem that were simply indicated there. He is the energy of the electron, which is the sum of its kinetic energy which, non-relativistically, is written $p^2/2m$, $p$ being the momentum and of the potential energy U due to the interaction with the nucleus. We can leave $H_{em.}$ simply indicated and $H_{int.}$, in the case of pure radiation, reduces to $\mathbf{ev}/c\mathbf{A}$; so that (4) can be rewritten:

$$\mathrm{H} = \frac{p^2}{2\,m} + \mathrm{U} + \mathrm{H}_{\mathrm{em}} - e\frac{\mathbf{v}}{c}\ \mathrm{x}\,\mathbf{A} \tag{6}$$

If then, applying the usual methods of perturbation theory, we study the shift of the terms of the spectrum of the hydrogen atom due to the perturbation $\mathbf{ev}/c \times \mathbf{A}$, the result is as follows: the first approximation gives zero and the second approximation diverges. That is to say, of the first two terms of the expansion, the term proportional to $e$ (of the first approximation), has zero as coefficient; the term proportional to $e^2$ (of the second approximation), has one of the usual infinities as coefficient.

Usually at this point we were discouraged and said: evidently, the theory is not good enough to deal with these phenomena.

The very simple observation that came out of the conference I mentioned is as follows: let us consider the expression $p^2/2m$. What do we mean by the $m$ that appears there? Since the division of the Hamiltonian into three terms is based on the idea of separating quantities relating to the electron alone, quantities relating to the field alone and interaction, the quantity $p^2/2m$ must refer to the electron alone. That is to say, if one could, by some unimaginable trick, exclude the electromagnetic field the mass $m$ to be put in the denominator of $p^2/2m$ would be the mechanical mass. In doing the calculations, we put, however, instead of $m$ the physical mass of the electron, which is $m_f = m_{em} + m$ and who knows how much of this is mechanical mass and how much electromagnetic mass. This is clearly not correct. Since the physical mass, the one we measure, is the sum of the mechanical mass and the electromagnetic mass if we want to correctly rewrite expression (6) making use of what we know, namely the physical mass, we must write it differently and precisely:

$$\frac{p^2}{2(m_f - m_{\mathrm{em}})} + \mathrm{H}_{\mathrm{em}} - e\frac{\mathbf{v}}{c}\,\mathrm{x}\,\mathbf{A} \tag{7}$$

We must now remember that the electromagnetic mass, as I said before, is a second-order effect and that is proportional to $e^2$. Therefore, if we want to formally make an expansion consisting of powers of $e$ we must develop the term $p^2/2(m_f - m_{em})$ and associate the part of it that contains the electromagnetic mass with terms of order $e^2$. We must therefore rewrite (7) in the following way:

$$\frac{p^2}{2m_f} + \mathrm{H}_{\mathrm{em}} + \mathrm{U} + \frac{p^2}{2\,m_f^2}\left(\frac{m_{\mathrm{em}}}{e^2}\right)e^2 - e\frac{\mathbf{v}}{c}\ \mathrm{x}\,\mathbf{A}$$

where I divided and multiplied $p^2/2m_f^2 m_{em}$ by $e^2$ to remind me that it is of the order of $e^2$. So that we come to have, in a sense, two perturbations. One is the one we considered earlier $\mathbf{ev}/c \times \mathbf{A}$, which gives zero as result in the first approximation and a result proportional to $e^2$ with infinite coefficient in the second approximation; but we also have another second approximation term that comes directly from:

$$\frac{p^2}{2\,m_f^2}\left(\frac{m_{\mathrm{em}}}{e^2}\right)e^2.$$

In calculating the coefficient of the term in $e^2$ both of these perturbations intervene and both result in an infinite coefficient, (remember, as I said, that the electromagnetic mass contains that certain infinity).[2] The essential point then is that the two infinite parts cancel out; and what remains is a finite result that is extremely close to what is observed experimentally. In saying this I am not completely telling the truth in that Bethe's work is a nonrelativistic calculation. It is therefore a calculation that evidently falls away when we come to include energies of the order of $m_f c^2$. The result now said is true if to Bethe's work we add those of various other authors who have made it possible to study relativistic behavior; but of this I shall say something later.

Perhaps more by way of curiosity than anything else, I would like to show by a very simple argument due to Welton what is, or what can be considered to be, the mechanism of the fact that interaction with the electromagnetic field results in a shift of the S levels. Unfortunately, I do not have the time to carry out the argument in detail although it is an easy little calculation; I do not know whether I would believe the argument if it did not give the right result but in fact it does.

It comes down to this: let us consider an electron. Even if there are no photons present, there are always the fluctuations of the electromagnetic field which produce the well-known fact that if we measure, for example, the electromagnetic field at a certain position, in reality, even if there are no photons we find a value other than zero, due to these fluctuations. Now the fluctuation field shakes the electron by forcing it, if the electron wanted to be at rest, to move with some kind of forced oscillations. Thus the electron spreads itself over an area of finite extent; this alters the interaction between the electron and the charge of the nucleus, which we can consider point-like, in the sense that when the electron comes very close to the nucleus we have to take this extent into account; and, if we do the calculation on this basis we find precisely the same result as obtained by Bethe. Thus we can say that the effect is due to fluctuations in the electromagnetic field, which is also the point of view arrived at by the more complete theory.

Let us say a few things about this more comprehensive theory.

It has been attacked by various researchers, and perhaps the people who have contributed most to this research are Schwinger, Tomonaga, Feynman and Dyson. I will not go into what each of these persons has done. They have contributed in various directions and in various limits.

The problem solved by Bethe is a nonrelativistic problem; and in fact the solution is really quite incomplete, as I have already mentioned, because, in order to bring it to convergence, one has to truncate the process at energies of the order $m_f c^2$, where the electron begins to behave non relativistically, which is in a sense justified. On the other hand, without a thorough analysis one could not know what happens at that point.

---

[2] The "type" of infinity depends on the type of wave equation the electron obeys.

The problem therefore arises of establishing a methodology for making such calculations which is relativistically correct; and this requires in particular that we write the equations, shall we say, of motion of our system, in a form that is clearly relativistically invariant.

In all common representations of quantum field mechanics this relativistic invariance, although it exists, as can be seen through complicated calculations, is usually not very evident for the simple reason that the Schrödinger equation, for example, represents a law of state variation as time varies, and consequently in using the Schrödinger equation one is dealing with variations with respect to sections of space-time at constant time. And this procedure is certainly not the best one for showing the relativistic covariance of a theory.

Tomonaga's idea was to develop a method for dealing with field theories in which consideration of the state at a certain time disappears. One might naturally think of proceeding in the following way: we consider the surface families $t' = \text{cost}$ "parallel" to the space of a generic laboratory next to the surfaces $t = \text{cost.}$ "parallel" to the space of our laboratory, and see if it is possible for us to try to establish, given the state on one of these sections, equations that would allow us to work on the other sections.

The procedure just mentioned would be a relativistically invariant procedure; but it seems so far that it would be mathematically a little more complicated than the following more general procedure due to Tomonaga. Instead of considering flat sections of space-time let us consider any sections; in other words, let us consider an arbitrary surface, with the only restriction that it be a spatial surface, i.e., that given any two points on this surface, their distance in time, multiplied by c, be less than the distance in space, which property is known to be a relativistically invariant property. Given any spatial surface $\sigma$ one can search for an equation that allows us to determine the state itself on a $\sigma^{'} \neq \sigma$, given the state at all points of $\sigma$.

And to be precise the question should be posed in the following terms: let us fix our attention, indeed, on the Schrödinger equation which, as is well known, tells us how the state varies in the transition from one to another of the $t = \text{cost}$ surfaces "parallel" to the space of our laboratory. Schrödinger's equation is famously written:

$$\mathrm{H}\psi = i\hbar\frac{\partial\psi}{\partial t} \tag{8}$$

$\Psi$ being the state function of our system and H being its Hamiltonian, which, in field theory, always occurs as the volume integral of a certain Hamiltonian density; I will denote the latter by $\mathscr{H}$ therefore $\mathrm{H}=\int \mathscr{H} dV$.

So, in accordance with the ideas discussed above, it is a question of finding an equation of a functional character that allows one to associate a state function not with a time, as in (8), but with a surface of spatial character $\sigma$, and that enjoys the following properties. First, such a functional equation must reduce to the Schrödinger equation if the surface succession being considered is a $t = \text{const}$ surface succession. Second, if one goes from an initial surface $\sigma$ to a final surface $\sigma'$ in various ways, passing through intermediate surfaces of various shapes, the result must depend only on the specification of the initial surface and the final surface.[3]

This leads to the solution of this problem, to the Tomonaga equation, which is usually written in the following notation that I will explain now

$$i\hbar c\frac{\delta\psi(\sigma)}{\delta\sigma(x')} = \mathcal{H}(x')\psi(\sigma) \cdot \tag{9}$$

In (9) in place of the time parameter the surface $\sigma$ which is our parameter appears as the argument of $\psi$. The point $x'$ is a four-dimensional point on this surface and the symbol $\delta/\delta\sigma(x')$ (functional derivative with respect to $\sigma$ of the point $x'$) is there to remind us that $\psi(\sigma)$ is a functional of the surface parameter. The functional derivative should be understood as follows: suppose we change the surface $\sigma$ slightly in the vicinity of the point $x'$. This will lead to an infinitesimal change in $\psi$, let us say $\psi(\sigma)$ the value of $\psi$ relative to the surface $\sigma$, $\psi(\sigma + \delta\sigma)$ the value of $\psi$ relative to the surface $\sigma$, $\psi(\sigma + \delta\sigma(x'))$. Then the functional derivative is defined as the variation of $\psi$ divided by the four-dimensional volume that has been added, when the latter tends to zero by continuing to contain the point $x'$

$$\frac{\delta\psi(\sigma)}{\delta\sigma(\mathrm{x}')} = \lim_{\delta\sigma(\mathrm{x}')\to 0} \frac{\psi(\sigma + \delta\sigma) - \psi(\sigma)}{\int \delta\sigma(\mathrm{x}')d\mathrm{V}}$$

It can be shown very easily that this Tomonaga equation has precisely the required properties, provided there is a relation that is always verified in the cases that arise, namely that the Hamiltonian densities taken at two four-dimensional points, that are however connected by a space vector, commute:

$$\left[\mathcal{H}(x),\ \mathcal{H}(x')\right] = 0. \tag{10}$$

The Tomonaga equation has the great advantage of being relativistically invariant at first glance and is therefore far superior for practical calculation to the Schrödinger equation whenever one has to be careful about relativistic invariance, since this invariance can be seen directly without having to prove it.

Having put the formalism on a relativistically invariant basis, it is certainly no easy feat to say how one proceeds in Schwinger's method in the short time I have left! I will mention the essential points. In Schwinger's method we still start from an expression of type (4) and perform a series expansion in powers of e. The problem is treated here directly with an equation or equations that are of the Tomonaga type and one naturally finds divergences, i.e. terms represented by integrals, that are divergent.

[3] In fact, the magnitude H that appears in (9) [and in (10) below] is not really the previously defined Hamiltonian density but the so-called interaction Hamiltonian density.

An attempt is then made to group these terms conveniently. It is precisely seen that:

1st some of them have the character of mass. Essentially they are terms that represent electromagnetic mass and they are associated with mechanical mass in giving the total mass without one bothering that they are infinite. It is said that the only quantity that we can know that has mass character is the total mass, the physical mass, and dividing that mass into a mechanical mass and an electromagnetic mass probably does not make sense. Or at any rate we would not know how to do it even if we wanted to;

2nd other terms have the nature of an electrical charge and are associated with it giving a "renormalization" of the charge.

I cannot insist much on this point. I will simply say that just as we have introduced an electromagnetic mass, a kind of, let us say, electromagnetic charge, which is connected with the so-called phenomenon of the polarization of the vacuum, is also introduced in Dirac's theory of the electron. In Dirac's theory the vacuum has a physical character, and it is polarizable, so that a charge, by the very nature of its field changes the vacuum and changes itself.

We then proceed here in a completely analogous way to what was done in the case of mass by saying: distinguishing an electric charge into electric charge, we can say, original and polarization electric charge makes no sense; the only quantity having the character of charge that we can know is the physical charge, the total charge: $e = 4.8 \times 10^{-10}$ u.e.s. And this quantity is the one we should substitute in place of all that set of terms having the character of charge without caring whether those terms are infinite.

3rd finally other terms of a different nature remain. They are also divergent, but divergent in an, as it were, indeterminate way, in the sense that they have positive and negative parts, so that they are either divergent or convergent depending on the order one chooses. Here an arrogant act is performed by placing all these terms equal to zero and justifying it in the following way. The scheme must of course satisfy the general conditions of invariance: relativistic invariance on the one hand and gauge-invariance on the other. The aforementioned terms always have a tendency to go against one or the other of these, being coefficients of non-invariant expressions. Since there is a possibility of making them equal to zero, the philosophy of the theory is to say that evidently in a complete theory they will have to prove to be zero because of the necessities of invariance. So we anticipate this final theory by placing them at zero by default.

What I have outlined is very briefly Schwinger's idea. Feynman's and Dyson's ideas are somewhat different but not essentially distinct, at least up to the second approximation. In a sense they are better suited to dealing with collision problems and have the advantage that one can prove their uniqueness in all approximations so that perhaps they are indeed the answer to the problem of quantum electrodynamics.

I must say that this answer is apparently limited presently to the study of electrons obeying the Dirac equation. In the case of different particles the divergences could not be eliminated by the simple artifice I have indicated but would remain in essential form in the theory.

Finally, regarding the experimental results, I wish to add that in addition to Lamb's mentioned result, there is another experimental finding that has been interpreted in this same theoretical scheme, and that is a slight difference between the experimental magnetic moment of the electron and Bohr's magneton. It is a small difference of about one part in a thousand in that the experimental magnetic moment of the electron is seen to be equal to the following quantity:

Experimental magnetic moment = $(1+1/2\pi\ 1/137)$ Bohr's magnetons but, of course in a quantity as fundamental as the magnetic moment of the electron, it is not a negligible difference, and even this is essentially due to coupling with the electromagnetic field.

So we can conclude this discussion in the following way: in electrodynamics as such there has certainly been great progress. It is probably not yet the ultimate theory, but it has brought us closer and closer to understanding the phenomena. There is some hope that ideas of this kind can also be applied, for example, to mesonic theories, which are far more unknown than the electromagnetic field and consequently arouse much more curiosity at the moment. There are serious difficulties and it is by no means certain and perhaps not even particularly likely that these attempts will be very successful. Clearly, however, they are a field of future research of great interest. It may be indeed that the physics of the phenomenon is still so imperfectly understood that the attempt to make too detailed a mathematics of it is necessarily doomed to failure. Or it may be that mathematics will allow the true physics of the phenomenon to be understood. This cannot be said at the present time.

What I can say is that there is at this time a great deal of activity. Almost all young theoretical physicists are eager to deal with field theories and nothing but field theories, perhaps even a little too one-sidedly: but only the future will tell.

I want to conclude by thanking again the audience who had the patience to follow me through sometimes abstruse and sometimes boring considerations. And I want to thank again the Donegani Foundation and the Accademia dei Lincei for giving me the opportunity to spend these two weeks at the University of Rome.

## Biographical Notices on Giacomo Morpurgo

### AT and PT

Giacomo Morpurgo was born in Florence in 1927. He graduated in Physics at the University of Rome in 1948. He was professor in Parma, Florence, and finally in Genoa. His main interest was in elementary particle physics. In this topic he collaborated with Luigi A. Radicati di Brozolo (1919–2019), Marcello Cini (1923–2012), Bruno Touschek (1921–1978). He is best known for his contribution to the non-relativistic quark model. He was a member of the CERN policy committee in 1982–88. He created an experiment for the search for free quarks using the magnetic levitation technique. He was the author of the Introduzione alla Fisica delle particelle in 1987.

For further informations see Morpurgo (1987), Salvini (1993).

## References

Morpurgo, Giacomo (1987). Introduzione alla Fisica delle Particelle (Bologna: Zanichelli, 1987)

Salvini, Pietro (1993). "Morpurgo, Giacomo" (Roma: Istituto della Enciclopedia italiana, 1993). 5th Appendice 1993 https://www.treccani.it/enciclopedia/giacomo-morpurgo/ accessed in September 26, 2023

# Seventh Lecture The Neutron

**Enrico Fermi**

## Edited by C. Salvetti

I want to thank Prof. Castelnuovo for his kind words toward me, and my gratitude goes to the Accademia dei Lincei and the Donegani Foundation which allow me to be here among you to speak to you about the studies to which I have devoted myself for many years. I am particularly pleased that this lecture is taking place in the hall of one of Italy's great industries. It is often said, and rightly so, that science contributes to industry. In a sense all great industrial developments have their root in one or more scientific discoveries. Less mention is perhaps made of another fact which is equally true and which is the opposite of what I have said and precisely how much science benefits from industrial progress. No modern science would be possible without the aid and contribution of those technical means which only great industrial development makes possible.

I plan to give three lectures here in Milan, of which this is the first. This one and the next will be devoted to problems in neutron physics, and this first one will be rather general and introductory; the following one will be more specialized. The third lecture on Dirac's magnetic monopole (a particle that probably does not exist ... but if it did exist it would be beautiful!) will also be specialized.

The neutron was discovered in the years 1932 and 1933 by a collaboration beginning with the German Bothe, continuing with the French scientists Joliot-Curie, and ending with Chadwick in England. It is a relatively heavy particle, it has the mass of one unit of atomic weight, as you know, and has zero charge, which gives it its name. Its great importance in the structure of matter was recognized almost immediately after its discovery, due mainly to Heisenberg, who recognized that the neutron is one

(*) Held on October 18, 1949 in the assembly hall of the Montecatini Company in Milan.

E. Fermi (✉)
University of Chicago, Chicago, IL, USA



A. Treves and P. Tucci (eds.), *Enrico Fermi, Atomic Physics Lectures*, History of Physics,
https://doi.org/10.1007/978-3-031-68690-0_10

of the two particles that form the atomic nucleus. However, atomic nuclei, as you have certainly heard many times, are made up of varying numbers of neutrons and protons which determine their charge and mass. In briefly outlining the history of the neutron, I apologize if I will consider it from my own personal point of view, and this is because I have already worked with neutrons for so many years by now that I consider them to be somewhat my relatives.

In 1934 the Joliot-Curies, the husband and wife couple whose research I mentioned a moment ago, made another very important discovery. By bombarding a number of substances with alpha particles they discovered artificial radioactivity, that is to say, the bombarded substances converted into radioactive substances. The discovery was extremely important being the first case in which radioactivity escaped those certain laws of ineluctability because of which it was believed until then that radioactive phenomena could not be produced in any way. After the discoveries of the Joliot-Curie couple it was learned for the first time how to manufacture radioactive substances.

**Fig.1**

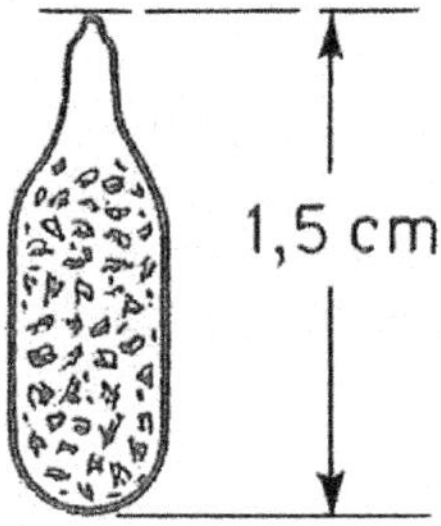


Now my first personal contacts with the neutron originated from this discovery of the Joliot-Curie couple. A rather obvious idea was that not only alpha particles could produce this artificial radioactivity, but probably neutrons as well. Indeed, one might have assumed, what experiment later confirmed, that neutrons must have certain advantages over alpha particles because, being electrically neutral, they would be able to approach the positive nucleus without being repelled by it. But neutrons are a little more difficult to produce than alpha particles because there are no direct sources of neutrons, but they are themselves the product of nuclear disintegration. So a source of neutrons had to be found.

At that time, at the Institute of Physics in Rome where I was working in those years, there was a fortunate circumstance in that Prof. Trabacchi, director of the Institute of Public Health Physics worked in the same Institute. He had at his disposal an apparatus for extracting radium emanation: and it was mainly with Trabacchi's help, and because of Trabacchi's help, that it was possible for me to obtain neutron sources, which were extremely small in comparison with the scale of neutron sources that can be available today, but entirely sufficient for a considerable amount of research.

I can briefly describe how these very simple sources were made. One of them was about 1.5 cm in size. It was a small glass tube of the shape shown in Fig. 1 in which beryllium granules were placed. Before sealing the glass tube a certain amount of radium emanation from Trabacchi's apparatus, for example half a Curie, was introduced into it. The alpha particles emitted by radon collided in large numbers with the beryllium and produced neutrons there, so that a large number of neutrons were released from this small source. A source such as I have described produces several million of them per second. This, of course, is not the only nor the most perfected neutron source; it is perhaps the smallest and is certainly one of the most convenient because it is a maintenance-free source. Once the emanation is introduced into the tube it continues to emit alpha particles on its own without needing any intervention or help from the operator. It is also one of those sources that never fails, which is also an advantage.

But then of course there are many other and much more powerful neutron sources, the cyclotrons and all the various machines that produce ion acceleration at energies of the order of millions of volts. These machines can be used as neutron sources by dropping the accelerated particles on top of a target of, for example, beryllium. Essentially it is the same principle only instead of using natural alpha particles one uses artificially accelerated particles. These are sources that generally give much larger intensities but are extremely complicated given the size and cost of the accelerator machines.

Today, as I will say in a moment, one of the most convenient, but also rather large and cumbersome, sources of neutrons is the pile in which uranium chain reactions are produced and in which intensities of an order of magnitude several times greater than what can be achieved with a cyclotron are reached.

These are, in brief, some of the major neutron sources.

Now to produce artificial radioactivity by means of neutrons one proceeds experimentally in a very simple way if one has one of these small sources available. To be precise the experiment is done in the following way: around the neutron source in its immediate vicinity one places, for example, an aluminum plate, an iron plate or a plate of the element one wishes to study and leaves it for a certain time which may be minutes, hours, days, as the case may be. The neutrons that come out of the source hit some of the nuclei of this substance; a variety of reactions then take place, which can be of very different kinds and which can roughly, be described as follows. When the neutron approaches a nucleus of the bombarded element it is not repelled because it is neutral. The probability of reaching the nucleus is therefore quite high. Of course, the nucleus is small and therefore difficult to hit, but the nucleus does not defend itself, so to speak. When the neutron hits the nucleus several phenomena can occur. It may be that the neutron is simply absorbed by the nucleus, which is thus excited. It may emit electromagnetic radiation in the form of gamma rays and transform into a nucleus whose weight has increased by one unit. This reaction is usually denoted by the symbol $(\nu, \gamma)$. The other most common reactions are: the reaction that is denoted by $(n, p)$ which means that the neutron enters and drives out a proton; the reaction $(n, \alpha)$ in which a neutron enters, and an alpha particle leaves. And there are others, but I will not mention them for the moment.

In each of these reactions a different nucleus is formed from the affected nucleus because in one case a neutron has been added and in the other cases examined a neutron has been exchanged for a proton or an alpha particle. It usually starts from a stable nucleus and ends up most often in a non-stable nucleus, that is, a radioactive nucleus. Now radioactivity can be measured by experimental methods of such refined sensitivity, the Geiger and Müller counters in particular, that even when a few radioactive atoms disintegrate every second or every minute, the counter manages to detect them. This is why the phenomenon, although quantitatively very small is actually very easily observed. I must confess that until that time, and to a large extent even afterwards, I was predominantly a theoretical physicist, and if I have sometimes succeeded in carrying out extremely simple experiments, I must say that as soon as the experiments become a little complicated they are beyond my experimental ability.

Fortunately at that time at the University of Rome I was working in collaboration with a group of students, later professors, and now finally colleagues. There was Amaldi, who is now director of the Physics Institute in Rome; there was D'Agostino, who was the chemist of the group (in research of this kind there is a need for a lot of chemical work, as I will say in a moment); there were Pontecorvo, Rasetti and Segré. Everyone was working at that time under the paternal direction of Prof. Corbino, who was then director of the Physics Institute in Rome. So what I did not know how to do in this experimental technique these friends of mine did know. I have mentioned chemistry, and I would like to indicate right now what particular chemistry was needed. Take, for example, the case of the reaction $Fe_{26}$ $(n, p)$ $Mn_{26}$. If the iron nucleus that contains 26 protons (because it has charge 26) loses one, exchanging it with a neutron that is neutral it loses one of the charges and is left with 25: this is the atomic number not of iron but of manganese. If you start with iron of atomic weight 56 you form manganese of weight 56 which is a radioactive manganese of an average lifetime of two and a half hours. Now if you do not know beforehand what reaction is supposed to take place, only chemistry can help you understand what happened. In iron bombarded by neutrons, a number, perhaps of the order of a few million, of radioactive manganese atoms are formed here and there. And how do we know that it is indeed manganese?

At this point our friend D'Agostino would pull out his test tubes and chemistry apparatus and performed, essentially, a separation of manganese from iron. For such an operation a small trace of manganese is usually added because the few manganese atoms that have been formed by bombardment feel "comforted" and "guided", so to speak, by the presence of a likewise small, but increasing, number of true manganese atoms, common manganese. However, after the addition of the manganese the whole thing is dissolved, separated, and on one side goes the manganese fraction, on the other side goes the iron fraction. Now by simply bringing one or the other of these fractions close to the counter and observing where the activity comes from, one can decide whether the activity is in the iron, in the manganese or in some other thing.

This is the kind of chemical work that was being done in Rome, in those years; work conducted in series, to the point of doing chemical separations of this kind even in less than a minute when dealing with radioactive substances with very short average lifetimes in which there was no time to dwell on elaborate chemical separations. Now, as I said, neutrons are a very powerful means of producing artificial radioactivity, so much so that in the period of about a year about fifty or so new radioactive elements were identified, in the Physical Institute in Rome by this method. And at that time we were all satisfied with the work done and it was believed that neutrons had given more or less all they could give. But shortly thereafter our group had the good fortune to make one of what are called chance discoveries. It was an experience that frankly I never expected, because being a theoretical physicist I should be able to predict what is going to happen; but instead this accidental discovery happened by chance without anyone going looking for it, and I will tell you how it came to be that we discovered the so-called slow neutrons.

So at that time there was some confusion in the laboratory; one necessarily worked under conditions that were not perfectly reproducible because there were many elements to investigate and sometimes one had to investigate one and sometimes the other in a way that was not always identical in either case. Now, in this somewhat irregular way of proceeding one began to notice differences that seemed inexplicable, and that is, sometimes the activity induced under certain conditions turned out to be small, sometimes it turned out to be larger. It even happened that by placing the plate to be irradiated far enough away from the source, where the activity should have been almost unmeasurable, some activity was measured instead.

Since the phenomenon, though strange, could be reproduced, a series of observations began to be made somewhat at random, placing the objects to be activated now here and now there in the hope of catching some circumstance that would give a key to solving the problem; and the key came by chance when it happened that a small piece of paraffin wax was interposed between the source and the object to be bombarded. This small piece of paraffin wax immediately made the intensity increase, albeit by a little, and then the question arose: if a little paraffin wax does this, what will a lot do? A lot was put in, and a lot of paraffin wax actually produced a very large effect. The intensity turned out to be multiplied by a factor of the order of 20 or 30 or 50, so as to indicate that it was indeed a strange phenomenon. Then it was realized that this strange phenomenon was what a theoretical physicist should probably have expected: that is to say it was the phenomenon of neutron slowing. It occurs in paraffin wax due to the fact that this contains a large fraction of hydrogen.

The phenomenon, which in an entirely similar way is also produced by water, consists of this: the neutron source I have indicated—like indeed almost all neutron sources—emits neutrons with rather large energy, of the order generally of a million eV. But if these neutrons are emitted within a block of paraffin wax or within a tank of water, they collide with a hydrogen atom, and since the neutron and the hydrogen atom have, to a great approximation, the same mass, we are roughly in the case of one billiard ball colliding with another. The kinetic energy of the first one is divided into parts, on average equal between the bumping ball and the bumped ball, so that the bumping ball goes away with about half the energy involved. If it then undergoes a second collision, and then a third, and then a fourth, halving its energy in each of these processes, we have that, halved and halved, the energy is reduced to little. In the case of the neutron the energy would even go to zero if at some point thermal agitation did not intervene: That is, the hydrogen atoms in the water or paraffin wax are in thermal agitation, so that the phenomenon of slowing down does not go on indefinitely, but stops when the neutron has lost so much energy that it is in thermal equilibrium with the environment; so that a kind of neutron solution is ultimately formed in the water.

Of course, this is a very dilute solution, and also a very peculiar solution, because while normal solutions can be kept within a vessel, there are, on the other hand, unfortunately no vessels for maintaining neutrons. When the neutron reaches the walls of the vessel containing the water, it goes on and leaves. Thus it is a solution in a state of continuous outward diffusion and also a solution in which the dissolved substance continuously disappears because the neutrons combining with hydrogen are, albeit partially, captured. A neutron in water lives about 200 ms. This time is long enough to accumulate a solution that has very large activating properties. The activating property is very large because in very many nuclear reactions the cross section of slow neutrons is much greater than that of fast neutrons, that is, slow neutrons have a much greater probability than fast neutrons of being captured.

Thus these slow neutrons are fabricated in a very simple way. Then, of course, the problem of studying their properties arises: these properties have been studied, from the time of their discovery to the present day, by using means of ever-increasing power. The first experiments were made by the absorption method, selecting these slow neutrons. When we speak of slow neutrons, it should not be understood that they all have the same velocity; they all have a small velocity, but not the same; so they can be classified into bands, let us say, some somewhat faster, some less fast depending on the velocity and depending on the energy. Now the first indications of peculiar properties of these slow neutrons with regard to absorption were obtained from absorption experiments, which, however, I am not going to talk about now, for today these kinds of experiments are conducted by much more powerful methods. Methods have been found today to produce slow neutrons, but all of the same velocity, i.e., "monochromatic" neutrons, i.e., of the same energy.

There are now two methods for producing monochromatic neutrons. The first of these methods is based on the use of an artificial source, for example, the cyclotron. The method consists of this: in the cyclotron a current of ions is produced that hits the beryllium target from which at the moment of impact neutrons are born. But in a cyclotron it is possible to modulate the source in such a way that this bombardment does not take place continuously, but occurs only at certain determined intervals of time. This involves opening the source only for brief moments at regular time intervals, what can be achieved by an electrical modulation method. Intermittent beams of ions are then produced in the cyclotron which, hitting the beryllium plate, give rise to instantaneous neutron bursts with adjustable modulation at will. Now if these neutrons are slowed down, for example by paraffin wax, and then sent to a detector placed at a certain distance, it is clear that of the neutrons, all slow, coming out of the paraffin wax the less slow ones will arrive on the detector first, the slower ones last. Now, if one selects by electrical methods those that arrive in a certain time interval, one selects neutrons belonging to a certain velocity band. That band can be chosen at will simply by shifting the time interval between the time of emission and the time of arrival. This

**Fig. 2**

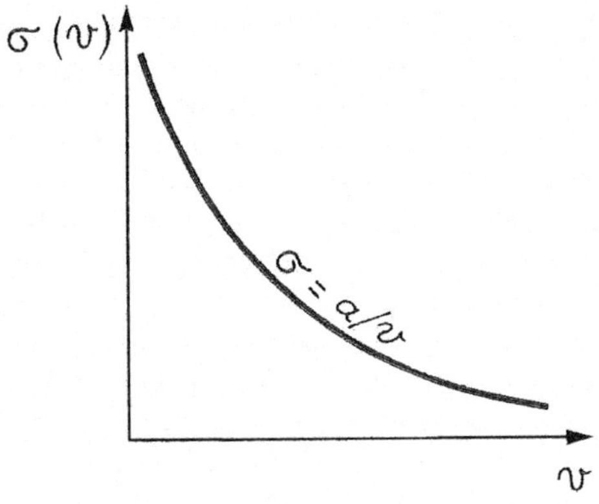


is one of the methods (Fig. 2).

**Fig. 3**

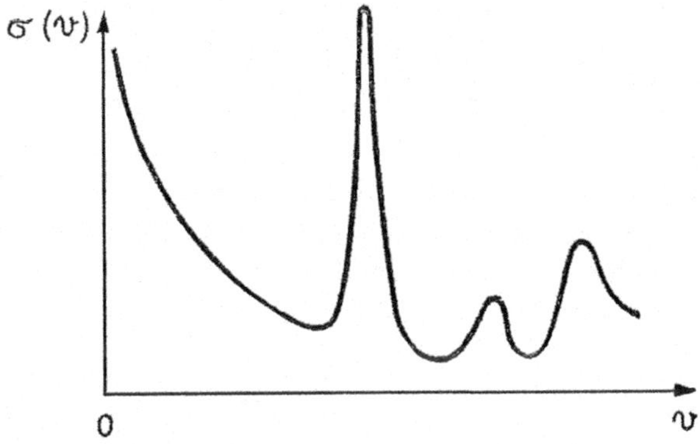


The other method will be described later: it is based on the use of atomic piles. These make it possible to achieve large neutron intensities, so that neutrons can be reflected on a crystal. Monochromatic neutrons are thus produced by the same method by which monochromatic X-rays are produced by means of crystal diffraction.

Either of these methods can be used to explore the law of the absorption of neutrons, a law that is sometimes quite simple: in fact, in some cases the cross section, $\sigma$, is a simple function of the velocity of the neutrons and to be precise is inversely proportional to their velocity. That is, the cross section given by $\sigma(v) = a/v$ ($1/v$ law). This is not only the simplest law of absorption but also indicates the reason why slow neutrons of small velocity are more active than fast neutrons; in other cases, actually quite often, the law is more complicated. One finds, for example, cases of neutron-absorbing elements with a law that can be represented graphically as in Fig. 3.

In general, for small velocities the $1/v$ law applies, but at higher velocities the cross section presents sometimes very pronounced and narrow maxima (resonances), which follow one other in a rather complicated way: this is the phenomenon of nuclear resonance, a phenomenon about which unfortunately I do not have time to give much information, but which serves very much to indicate certain peculiarities of nuclei that would be difficult to observe in any other way.

In the Physical Institute in Rome we had no great means, of course we had no atomic piles and no cyclotron. Only later, after I had left the Institute, was an accelerator machine built at the Institute of Public Health. But in those days there were no such machines, so studies of this kind could only be done by the very rudimentary methods of absorption. These methods are based on the property of selective absorption at the resonance lines. Now at the beginning of the study of these phenomena, for a few years, there was an attempt to understand at least some of the peculiarities of these phenomena, which were later clarified much more completely by experiments of another kind. After these experiments I thought that neutrons had more or less given all they could give and that I could peacefully return to my studies of theoretical physics, but another setback came with the discovery of splitting.

**Fig.4**

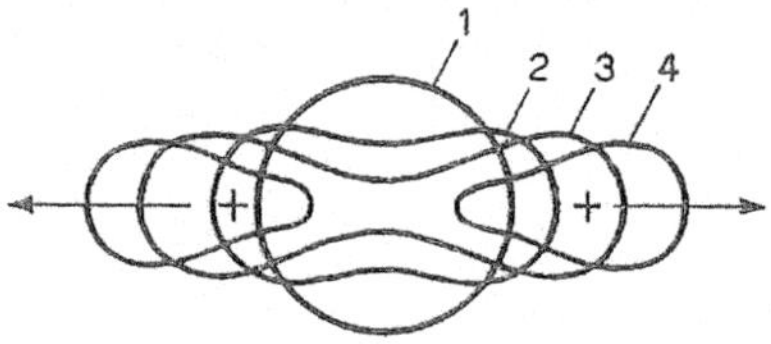


As early as my time in Rome, uranium had been among the other elements bombarded with neutrons. It had been noticed that uranium behaved in a strange way, but we had not realized how singular its behavior really was. In fact, bombarding uranium produced numerous activities. An attempt was made, with our rather primitive chemistry, to identify which elements were responsible for these activities, but it failed. It was found that such activity was always in elements other than those being examined. One examined uranium, protoactinium, thorium and down through the periodic system to, I think, radon, without ever finding the activity sought. Then it was thought that transuranic elements were formed. Actually transuranic elements do form, as you no doubt know, but in a very different way and under very different conditions than was then thought.

The solution of this strange behavior of uranium was found by Hahn and Strassmann, two German physicochemists, who, after many mistakes, because they too for a great while believed that they were repeating experiments of the kind made in Rome, realized that the error of our experiments consisted in having neglected to continue the investigation of the elements toward the zone of average mean atomic weights. These authors found precisely that elements located toward the middle of the periodic system are formed by the well-known phenomenon of uranium splitting, roughly as follows: a uranium nucleus, understood to be of a spherical shape, struck by a neutron goes into vibration; since it is already unstable in itself it happens that it sometimes leaves the position of equilibrium, stretching to the point where the recall forces are no longer capable of reconstituting its spherical shape. It follows that it goes through a whole series of configurations (represented in Fig. 4) until it splits into two parts.

The two fragments formed in this way are roughly equal: generally one is a little heavier than the other, although both are of the same order of magnitude. The two radioactive fragments formed speed away in opposite directions. They are in fact both positive, and as soon as the neck that unites them breaks, electrostatic repulsion causes them to move away from each other with enormous energy; this is the essential reason for the great amount of energy that is released in the splitting.

The splitting phenomenon came to my attention in January 1939. I had just arrived in New York and was somewhat looking around for a field in which to work when news of this discovery came from Germany. Naturally there was a period of excitement in physics: in a dozen laboratories, at the same time, experimenters were rushing to investigate the peculiar properties of this new type of nuclear disintegration. It was then that I realized for the first time that such a phenomenon might one day bring nuclear physics out of the narrow field of pure research and transport it into the field of "big things." I then had the idea confirmed by later experiments, that at the splitting of the nucleus into two fragments neutrons could also be emitted, just as when a drop is broken up one can sometimes see minute droplets spurting out. As a consequence it could be the case, for example, that the first neutron that produced the split produced on average, let us suppose, two: if this were so, of course, there would be the possibility of a chain reaction in which, from one original neutron, an arbitrarily large number could be produced by successive doubling, as long as there remained uranium to be split.

These ideas were much discussed, at a scientific meeting that was held in Washington in January 1939. Then another phase of my relationship with neutrons began in which I sought confirmation of this hypothesis. Experimental demonstration was brought by Joliot with some of his collaborators, by a group working with me at Columbia University and, more or less independently, by Szilard and collaborators. From then on, for several years, I devoted all my activity to experiments of this kind to see how far we could go. All this was happening at an extremely tense political time. War had not yet broken out, but it was already very much in the air and it was easy to foretell that it would be a ruthless war in which civilization itself would be at stake. This research, contrary to scientific tradition, was conducted partly in secret, and this was done, even before the government, military authorities or others became interested in the problem.

A group of physicists, part American, part European, living in America at that time, privately agreed to conduct this research with a certain confidentiality. So it went on for a number of years at first without, then with limited help from the American government, and finally with help that eventually reached spectacular figures. For that matter, you know the stages through which it went. In particular I want to mention here that, already at a relatively early stage of this research, a chain reaction was produced in a machine in which the neutrons are slowed down by graphite. This machine, the atomic pile, consisting of a large mass of graphite in which pieces of uranium are arranged with a suitable geometry, when it has reached a sufficiently large size, it begins to work spontaneously. The reaction is perfectly adjustable because if the intensity grows it is enough to introduce neutron-absorbing bodies. On the other hand, if you want to increase the activity level, you remove these absorbers so that you have an extremely quiet, adjustable, nonexplosive device.

But I do not want to speak to you about the atomic pile, neither as an instrument of a truly industrial character, nor as a prototype of those reactions which later led to the atomic bomb. Rather I would like to speak to you about the atomic pile as a means of research, a means which is indeed powerful, which will have, I believe, in the long run, a very important influence on the development of science. First I would like to mention the scientific applications of the pile as a neutron source for physics experiments. As a neutron source it has a great number of advantages, accompanied by a number of drawbacks. The main advantage is that it provides enormous intensities, far greater than those obtainable by any other means, and at the same time it is very constant and perfectly adjustable.

The main drawback is that very few laboratories have them. Fortunately, we had two in Chicago, so that a number of research projects were organized, many of which relate to the so-called optical properties of neutrons. These, like all corpuscles, obey quantum mechanics and therefore have the dual behavior of particles and waves and, when their wave-like appearance is experienced, they refract, reflect and exhibit properties similar to those of X-rays.

Another type of application is this: an atomic pile is a source of radioactive substances of great variety and intensity. These radioactive substances are produced in two different ways. First, the very phenomenon of splitting, on which the operation of the pile is based, leads to the production of radioactive substances (fission products); one can then, by appropriate and sometimes complicated chemical methods, extract, isolate, purify, and use these products for the purposes we shall see.

The other way to produce radioactive substances is to introduce into the pile a substance that absorbs neutrons, for example according to the $(n, \gamma)$ reaction, and turns into a radioactive substance. Given the wide variety of elements that can be introduced, it is understandable how a large number of radioactive isotopes can be produced. In the United States the production and distribution of radioactive isotopes is currently done, in a very systematic way, with an industrial organization using the Oak Ridge pile. The U.S. Atomic Energy Commission then puts them on the market at a nominal price, which is, I think, roughly one-twentieth of the cost price, available to laboratories that have a justified need for them. Isotopes serve a wide variety of research, in chemistry, biology (as indicators), and medicine, including as radium substitutes. Of course, both the manufacture and distribution of these isotopes present problems that are not simple. For example, shipping large quantities of radioactive substances cannot be done obviously by just wrapping them in a small package and putting them in a mailbox, but it requires certain precautions because of the activity that can be dangerous. So for example, a quantity of the order of a milligram of radioactive substance is shipped in a sizable lead envelope for protection against radiation, which increases its weight and cost. Requests for radioactive substances at Oak Ridge generally come from American and also European Institutes. Some shipments have also been made to Italy, for the Institute of Public Health in Rome. The shipment is often quite complicated, and naturally takes some time, so that short-lived substances cannot be shipped. If a substance has an average life of less than a few weeks it may decay before reaching its destination.

To conclude, I want to mention an application of radioactive substances produced by neutrons, which in this case are generated neither by the pile nor by any of the sources I have mentioned, but by the cosmos; that is, to the measurement of the age of mummies, historical or prehistoric relics that contain organic residues in any way. In the case of mummies the organics are contained in the cloths in which they were wrapped; in other cases beams of old dwellings, fossil residues, however preserved, etc., may be used, provided they contain some trace of carbon.

The really very ingenious method devised by Prof. Libby of the University of Chicago consists of this: cosmic rays, disintegrating some of the nuclei of the atmosphere produce neutrons in very small numbers, of the order of one per second for a column of air with a base of a square centimeter. Given the size of the Earth, however, the total number of neutrons produced in the atmosphere is considerable. They are absorbed by nitrogen, the only one among the constituents of the air that has considerable absorption for neutrons according to the reaction: ($N_{14}$+n = $C_{14}$ + $H_1$) by which one proton and one carbon 14 atom are obtained. Since this reaction takes place continuously in the atmosphere, on each column of a square centimeter of atmosphere, about one carbon 14 atom is formed every second, which is radioactive with a half-life of about 5.000 years; $C_{14}$ is the only radioactive isotope of carbon; the other two, $C_{12}$ (the common carbon) and $C_{13}$ are stable. The carbon 14 atom thus produced is rapidly oxidized, giving rise to carbon dioxide $C_{14}$ $O_2$ which, through chlorophyll function, rapidly becomes part of leaves, of plant matter. This is assimilated by animals, by humans, and on their death the carbon 14 returns to the atmosphere. That is to say, there is a continuous circulation of carbon 14 between the atmosphere and all living organic matter, whereby it is diluted with the carbon of the whole biosphere, so that each carbon specimen, obtained from living matter, will contain a percentage, naturally a very small one, of carbon 14. I will not go into all the experiments made and the difficulties Libby went through to make this phenomenon clear.

Among other things, he examined methane of two origins: one, very old, from oil wells, containing carbon that was originally part of living matter millions of years ago, but through the millennia in which it remained in the deposits it ended up losing all trace of the carbon 14 activity initially contained; the other, extracted from Baltimore sewers, therefore of recent organic formation, came from the cycle described. After a certain amount of research and refinement of technique Libby found that indeed the Baltimore methane was more active than the methane from natural gas wells and indicated the presence in the carbon of a minute but measurable amount of carbon 14.

And here is how this technique can be applied to find the age of a mummy. At the time of death, the body, which was then mummified, contained a certain percentage of radioactive carbon and therefore possessed a certain activity, but in the three to four thousand years since then, the carbon present in the mummy has decayed with a half-life of about 5.000 years. Therefore, if the mummy were, say, 5.000 years old, it would contain half of it today, if it were 10.000 years old, it would contain a quarter of it, and so on. By accurately measuring the specific carbon activity of the mummy, one can thus determine its age, to an approximation of about 200 years, which, at least in some cases, is, however, not great compared to the approximation of historians. In the case of Egyptian history, which is very well known in general, historians establish dates more accurately than physicists, but in several cases of history and in some of prehistory and recent geological epochs this method of deducing dates from radioactive carbon is of very great value, and is a valuable aid in some branches of knowledge which, one would think, would have little to expect from physics and particularly neutron physics.

I chose this example, and I could of course have chosen many others, to indicate how the study of radioactive substances, supported also by a certain amount of imagination, can find application in the most diverse fields of research, as well as in the more natural fields of chemistry, biology and medicine.

Thus I would like to conclude by expressing the hope that this neutron research, to which I have devoted so many years, will finally produce not only weapons, but above all a scientific development that can enrich humanity and extend the frontiers of human knowledge.

## Biographical Notices on Carlo Salvetti

### AT and PT

C. Salvetti was born in 1918 in Milan, where, in 1940, he graduated in Physics. There he was a student of Giovanni Gentile jr (1906–1942) who, in that period, taught Probability theory and Metal theory.

In 1941–43 he served in the Army. After 8 September 1943 he refused to join the Republic of Saló from where Mussolini hoped to reconstitute the Fascist regime and went into hiding.

At the end of the war in 1945 he began his research in the field of cosmic rays. Together with Giorgio Salvini (1920–2015), both assistants to Giuseppe Bolla (1901–1980), he was involved in the task of studying the possible developments of the peaceful use of nuclear energy. Mario Silvestri (1919–1994) and Salvetti drew up a research program for the construction of a zero-power atomic pile similar to Enrico Fermi's CP-1 in Chicago (1942) and the construction of a small power reactor.

The idea of setting up of a new center arose in this context.The new institution was founded in 1946 and was called Centro Italiano di Studi ed Esperienze CISE (Italian Centre for Studies and Experiments), a private research institute on nuclear physics and technology. In the early '50s he was a visiting Professor in India, assisting and guiding the Indian Atomic Authority in the development of a national chain of heavy water reactors.

In 1952 the Consiglio Nazionale delle Ricerche (CNR) (National Research Council) established the Comitato Nazionale per la Ricerca Nucleare (CNRN) (National

Committee for Nuclear Research) which in 1953 provided funding and personnel for the Frascati Electrosynchrotron, completed in 1959.

In 1956 the Ispra Research Centre was established, the first CNRN research infrastructure, and a 5MW CP5 experimental nuclear reactor was ordered. Salvetti was the director.

In 1960 the Italian Parliament transformed the CNRN into the Comitato Nazionale per l'Energia Nucleare (CNEN) (National Committee for Nuclear Energy). Felice Ippolito (1915–1997) continued to serve as Secretary General of CNEN. In the same year the Italian Government sold (on rather disadvantageous terms) the Ispra Research Centre to EURATOM and the CP5 reactor only a few months after its completion. The negotiation was conducted in a non transparent manner and Salvetti resigned.

In 1959 he became Director of Research at the Laboratories of the International Atomic Energy Agency (IAEA) in Vienna.

In 1963, following the defenestration of the director Ippolito for minor administrative irregularities, followed by a trial that imposed a heavy sentence, Salvetti was appointed vice-president of CNEN, replacing Ippolito.

He was appointed to a chair in Bari (1954), and later taught in Milan (1959) and in Rome in 1981.

He died in 2005.

For further informations see: Bonolis (2005); Gariboldi et al. (2022); Lavrencic and Valli (2005); Salvetti (1970).

## References

Bonolis, Luisa (2005). "Giovanni Gentile Jr a Milano" Atti del XXV Congresso Nazionale di Storia della Fisica e dell'Astronomia, Milano, 10-12 novembre 2005 (Milano: SISFA, 2005)

Gariboldi, Leonardo; Bonolis, Luisa; Testa, Antonella (2022). The Milan Institute of Physics. A Research Institute from Fascism to the Reconstruction (Cham: Springer International Publishing, 2022).

Lavrencic, Devana; Valli, Giulio (2005). "In ricordo del prof. Carlo Salvetti" Energia, Ambiente e Innovazione 2005 51: 23-24.

Salvetti, Carlo (1970). "Nuclear energy in Italy" https://inis.iaea.org/collection/NCLCollectionStore/_Public/01/000/1000475.pdf, accessed in September 28, 2023

# Eighth Lecture Optical Similarities in Neutron Properties

**Enrico Fermi**

## Edited by C. Salvetti

I would like to discuss this evening some aspects of neutron physics in which the wave-like character of these particles is highlighted. According to the results of quantum mechanics or wave mechanics, all particles give rise to interference phenomena, similar to those of light radiation, in which the wavelength is given by de Broglie's relation ($\lambda = h/\text{mv}$), that is to say the ratio of Planck's constant to the momentum mv of the particle. For such phenomena to be observed, it is necessary that the wavelength be of the order of magnitude of other geometric dimensions contained in the problem; it is therefore convenient to indicate without fail, its value. For so-called thermal neutrons, that is to say neutrons that are slowed down to thermal motion velocities corresponding, at room temperature, to the energy of about 1/40 eV, it is found from de Broglie's relation that the wavelength is about $1.8 \cdot 10^{-8}$ cm. Since this length is very close to interatomic distances, one expects to observe for neutrons diffraction phenomena in crystal lattices similar to the phenomena observed with X-rays, which also have wavelengths of this order of magnitude.

Now, before entering into the discussion of the various experiments that have been carried out to actually observe properties of this kind, I would like to establish some simple general facts and some general formulae that we will use later. Suppose that on an atom, or rather on the nucleus of an atom (because practically it is always the nucleus that produces the phenomenon of neutron scattering) a neutron arrives and hits it. The neutron will be scattered according to a certain angle (Fig. 1a). This is the pattern in the corpuscular representation. In the wave representation you have

(*) Held on October 20, 1949 in the assembly hall of the Montecatini Company in Milan.

E. Fermi (✉)
University of Chicago, Chicago, IL, USA

 
A. Treves and P. Tucci (eds.), *Enrico Fermi, Atomic Physics Lectures*, History of Physics,
https://doi.org/10.1007/978-3-031-68690-0_11

a group of incident waves which are scattered as light would be, that is to say in a group of spherical waves radiating from the scattering center (Fig. 1b).

**Fig.1** DIFFUSION OF THE NEUTRON INCIDENT ON A NUCLEUS

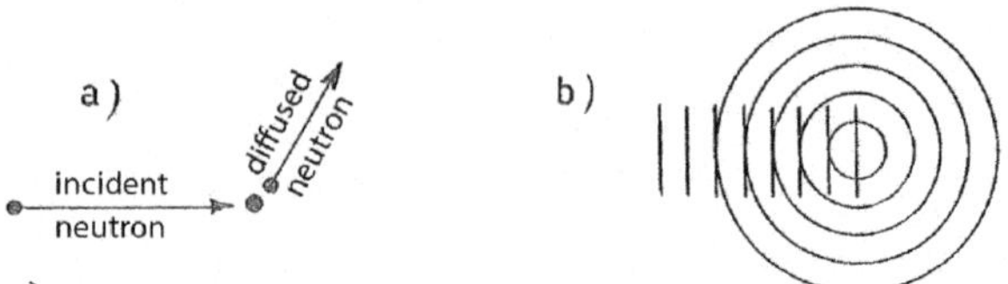


a) CORPUSCULAR REPRESENTATION

b) WAVE REPRESENTATION

Let the incident plane wave propagating along the positive direction of the x-axis be represented with the exponential

$$e^{ikx} \tag{1}$$

The scattered wave will be a spherical wave, which can be written

$$-a\frac{e^{ikr}}{r}\,. \tag{2}$$

The factor $ikr$ represents the phase (variable) of the spherical scattered wave. The factor $1/r$ indicates that the amplitude of the scattered wave decreases with distance from the scattering center by a law of inverse proportionality. This must be in accordance with the laws of conservation of energy, because the intensity of the scattered wave, which is proportional to the square of the amplitude, decreases as $1/r^2$.

By studying the scattered wave and following it in a given direction, one can ask whether or not there was a phase difference at the time of scattering. This question can be examined, at least in part, on the basis of some general principles on which, however, I will not dwell for reasons of time. Rather, I would like to state the result. Between the incident wave and the scattered wave there are generally small phase differences, but these are concentrated very close to the values 0° and 180°. In this scattering therefore either there is almost no phase change or there is a phase change corresponding to about a 180° phase inversion. As I repeat, I would not like to discuss in detail why only these two values are acceptable, however, it can be said that any phase change that was very different from either of these two values would correspond to a very strong absorption. The phase difference is determined by the constant $a$ that is involved in (2). To be precise, if the constant $a$ is real and negative, the phase difference is 0°; if $a$ is real and positive, the phase difference is 180°. Finally complex values of $a$ correspond to intermediate values of the phase difference.

I simply write down the relationship, which could be demonstrated very expeditiously, that links the constant $a$ with the cross section by diffusion:

$$\sigma = 4\pi a^2. \tag{3}$$

There is another point I would like to fix before discussing experimental work in detail, namely, the meaning of this constant $a$ in quantum mechanics, a meaning that can be illustrated in the following way.

Let us first observe that the wavelength is of the order of $10^{-8}$ cm, while the nuclear dimensions are of the order of $10^{-12}$ cm, and thus the wavelength is about 10 thousand times the size of the object producing the scattering. In the graph in Fig. 2 the distance $r$ from the center of the nucleus is shown in abscissae, the nucleus itself being able to be represented schematically by a potential well of a radius of the order of $10^{-12}$ cm, while in ordinates the product of r by the wave function is shown (in the case of energy $E = 0$). Starting from $r = 0$, the curve has a sinusoidal path up to the limit of the potential well representing the nucleus, and then acquires a straight path. Collision theory allows the constant $a$ to be given the following geometric meaning: it is equal to the distance from the origin of the intersection with the r-axis of the extension of the rectilinear part of the diagram. Figure 2 refers to the case $a>0$; for $a<0$ the intersection falls on the negative side of the r-axis (Fig. 3).

**Fig.2**

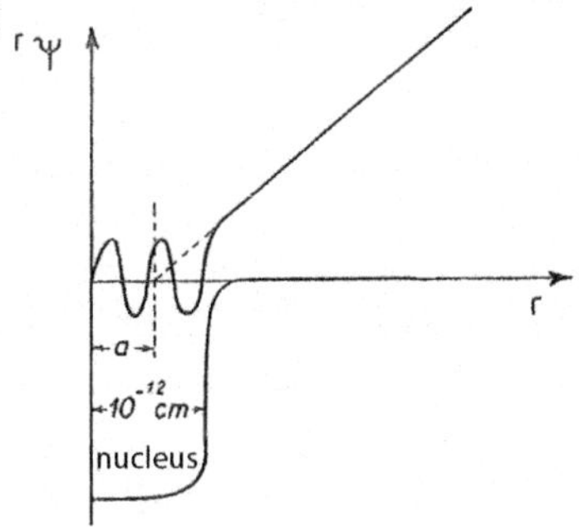


**Fig.3**

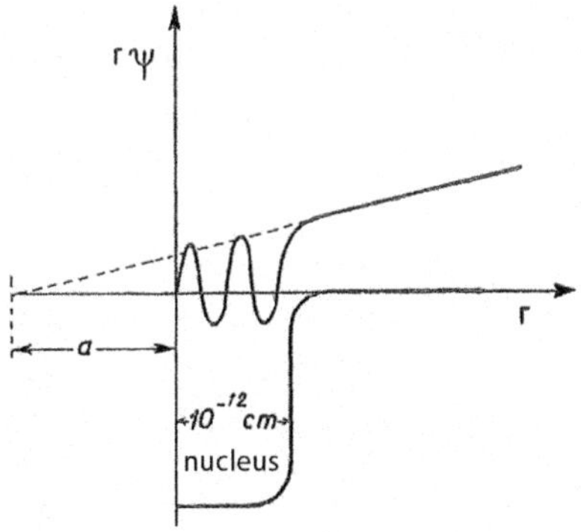


After these preliminary remarks, I would like to turn to a description of the simple experiment by which it can be verified that neutrons do indeed behave essentially like X-rays with respect to diffraction phenomena in crystals. If one lets monochromatic

X-rays fall on the face of a cut crystal, for example, parallel to a series of lattice planes (Fig. 4) one finds that reflection occurs only for certain special directions.

**Fig.4**

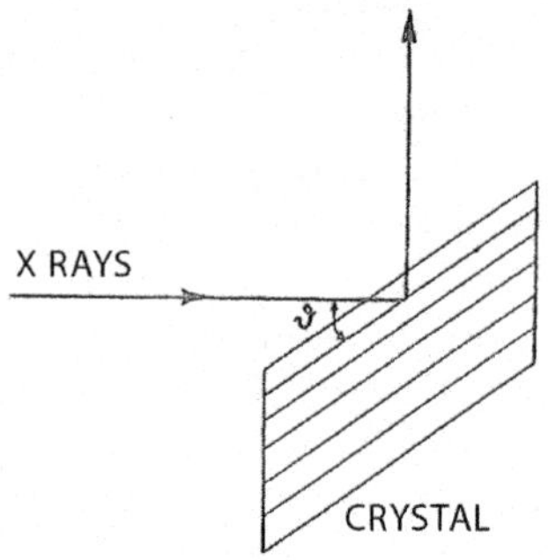


These are identified by the well-known Bragg relation:

$$n\lambda = 2d \sin \theta \tag{4}$$

in which, $n$ is an integer called the order of the reflection, $\lambda$ is the wavelength of the X-rays, $d$ the distance between successive crystal planes in the considered series, and $\theta$ the complement of the angle of incidence on the crystal. Now, if the general result of quantum mechanics is true, that neutrons behave both as particles and as radiation with wavelength $\lambda = h/\text{mv}$, we might expect a similar phenomenon to take place for neutrons.

This was precisely the purpose of the simple experiment that was first performed by Zinn. To place oneself in favorable conditions one must first have a very intense neutron source: very intense neutron fluxes occur in atomic piles, and therefore the device used by Zinn is the one depicted in Fig. 5.

**Fig.5**

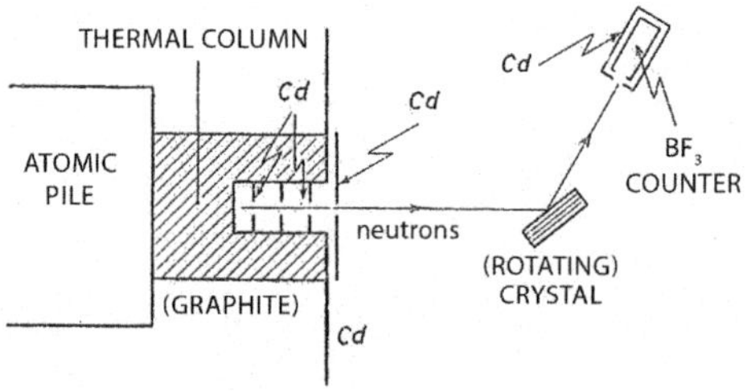


The pile is surrounded by a large screen, made of concrete for protection against radiation, into which, as in many piles built for research in the field of physics, is inserted the so-called "thermal column" that is to say a column of graphite, the end of which sinks into the pile, and which is intended to slow down the fast neutrons coming from the pile. They bump into the carbon nuclei in the thermal column and gradually lose their initial energy until they come into thermal equilibrium with the carbon nuclei in the graphite block outside. At the end of the graphite column a cavity is drilled, as the figure shows, in such a way that the thermal neutrons, coming from the bottom are "roughly" channeled outward, so that at the exit we get a (admittedly not too collimated) beam of neutrons, with energy distribution corresponding to the temperature of the thermal column.

But since for experiments of this kind, one needs a far greater collimation, that is, one needs to produce a fairly thin beam of neutrons, of fairly well-defined direction, certain limitations are introduced for the purpose of selecting only neutrons having a certain direction. For this, use is usually made of the property of cadmium of being a very strong absorber for thermal neutrons, so that a layer of cadmium, about half a millimeter or one millimeter thick, practically absorbs them completely. Then by interposing cadmium diaphragms in the cavity made in the thermal column, a fairly well collimated thermal neutron beam will be obtained. Then the experiment is conducted in a manner very similar to that described for X-rays, and that is to have the neutron beam incident over a crystal mounted on a turntable. At this point, however, it should be kept in mind that the neutrons coming out of the thermal column are not monochromatic. The velocity spectrum more or less corresponds to a Maxwellian distribution (Fig. 6).

**Fig.6**

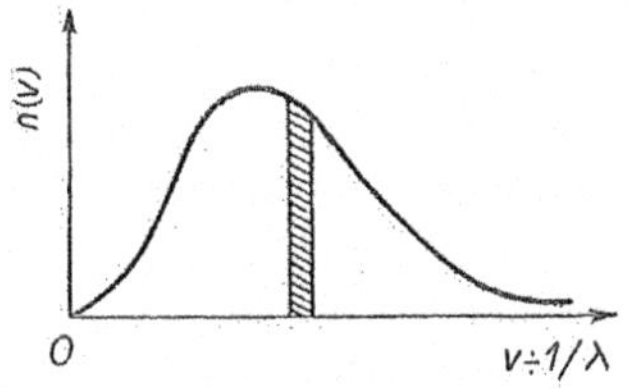


**Fig.7**

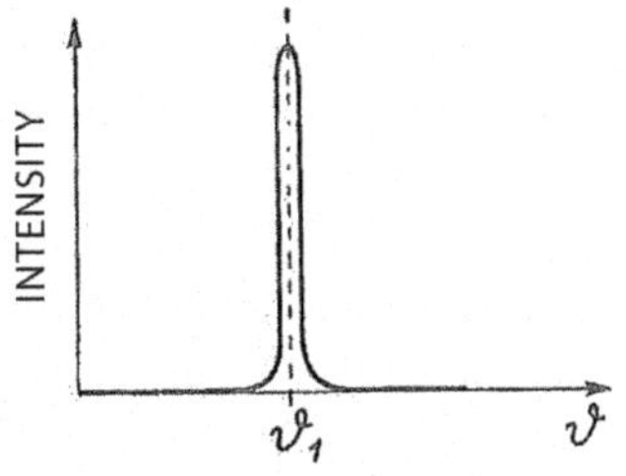


Since velocity is inversely proportional to wavelength, the x-axis in Fig. 6 is also essentially a scale of decreasing wavelengths, or rather, it is a scale of the inverses of wavelengths. So, of course, for every velocity interval, there corresponds a wavelength interval. Let us now give the crystal some specific orientation with respect to the incident beam. The Bragg condition will be verified only for a number of wavelengths, not necessarily for any one wavelength, because we have reflections of the various orders determined by the different values of the number $n$ appearing at the 1st member of (4). Fixing our attention for the moment on the value $n = 1$, of all the incident neutrons of various velocities only those of a rather narrow band are reflected, that is, those that have velocities and therefore wavelengths such that they satisfy Bragg's relation:

$$2d \sin \theta_1 = \lambda$$

The other neutrons in the beam largely traverse the crystal; therefore, in the direction identified by $\theta_1$, only a weak beam of neutrons, representing a small fraction of the incident neutrons, will come out; in other words, those neutrons that have a given wavelength were selected.

To observe reflected neutrons a neutron detector must be set up, mostly a boron trifluoride counter. To detect reflected neutrons according to a well-defined angle it is convenient to shield the counter with cadmium, fitted with a thin aperture. One can then perform experiments as one would with an X-ray tube, emitting a continuous spectrum. For example, if, keeping everything else fixed, one rotates the platform on which the crystal is placed, one generally finds very small intensities. When, however, the condition of angle of incidence equal to the angle of reflection is reached, the intensity suddenly rises very rapidly to considerable values. Increasing the angle of rotation again, the intensity again drops abruptly, so that if one represents the intensity as a function of the angle by which the crystal is rotated, one finds a diagram like that in Fig. 7 which has a very narrow and pronounced maximum.

By holding the crystal stationary at a position corresponding to the angle $\theta_1$ one can thus select from the continuous spectrum the neutrons of a small monochromatic band, corresponding to the Bragg conditions. From the earliest experiments in which these phenomena were brought to light, confirmation was sought that the wavelength thus obtained was indeed related to the crystallographic data and the geometrical data so as to satisfy Bragg's relation; and since $\lambda = h/\text{mv}$, an attempt was made to check the relation existing between the neutron velocity and the angle $\theta$. The simplest direct experiment for this purpose is of this kind: after making a measurement, with the crystal oriented according to an angle $\theta$ we rotate the counter by a certain angle. In order for neutrons to enter the counter thus rotated, we must also rotate the crystal, to a new position corresponding to a new angle $\theta_1$. In this way we can select a wavelength $\lambda_1$ other than $\lambda$, selected at the angle $\theta$. For example, if $\theta_1 < \theta$ the wavelength $\lambda_1$ is smaller than $\lambda$. As a result, a neutron band of different velocity is obtained.

How can one get to know the value of the neutron velocity in such cases? In other words: how can the previous relations be verified? The simplest way is to make use of an absorption phenomenon. In fact there are many substances—one of which is boron—that absorb neutrons with a cross section inversely proportional to the velocity. If one interposes, along the path of the neutrons, after they have been reflected, an absorber containing boron and thus measures the absorption coefficient, one finds that it is directly proportional to the sine of the angle $\theta$, as it must be precisely if the Bragg and de Broglie relations hold; for in such a case the absorption coefficient, having to be inversely proportional to the velocity v of the neutrons, will be directly proportional to their wavelength $\lambda$ and thus also to $\sin \theta$. Thus it can be verified that indeed Bragg's and de Broglie's relations hold for neutrons.

In experiments of this kind, one of the drawbacks is the use of non-monochromatic neutrons. Instead, more delicate experiments should be carried out using neutrons all of the same energy produced with appropriate devices.

The experiment is carried out roughly in the following way (see Fig. 8): one makes the beam of non-monochromatic neutrons whose velocity is given by the Maxwell distribution incident on a small crystal. In my experiment some years ago with Leona Marshall the crystal used was, for reasons of intensity, a fluorite crystal, but any other crystal will do more or less. Generally one leaves the crystal fixed in a certain position, although it is also mounted on a turntable, so that it selects a monochromatic band from the beam of primary neutrons, endowed with a continuous velocity distribution. Since in the process of reflection and monochromatization the intensity of the neutron beam decreases considerably, it is necessary to have a source of high intensity in order eventually to have monochromatic neutrons in sufficient numbers for the experiment. Along the path of the now monochromatized beam is then placed a second crystal, the one to be studied. Thus, the peculiarities of monochromatic neutron reflection by various crystals can be examined.

**Fig.8**

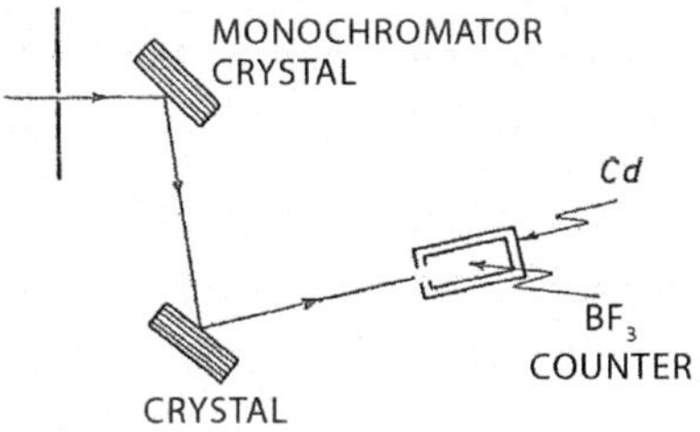


**Fig.9**

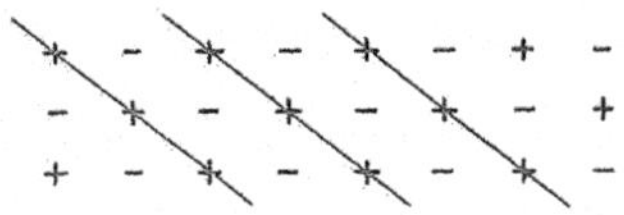


The beam reflected on the second crystal is revealed by a boron trifluoride counter fitted with a cadmium coating in which a slit is made. With experiments of this kind, systematic research was conducted on about thirty different crystals. The main purpose of this research was to clarify the question concerning the phase, that is whether a was negative or positive, whether the phase change was around 0° or around 180°.

I will cite later other methods of arriving at the same conclusion. This one I have described is based on the following argument: suppose, then, that we take a crystal, for example, sodium chloride, the structure of which is well known, it can be represented as in Fig. 9. In one plane there are alternating sodium and chlorine atoms, in the adjoining plane the same atoms are still alternating but the place of each sodium atom is taken by a chlorine atom and vice versa, and so on. If one considers the so-called face (100), which are ultimately the face of the crystal cube, one sees that the various reflecting planes are essentially identical since they all contain 50% chlorine atoms and 50% sodium atoms. If, on the other hand, we examine the diagonal face of the crystal, the so-called (III) face, we find that the reflective planes are different: in fact, the planes that are all made up of chlorine atoms are alternated with planes that are all made up of sodium atoms.

Suppose we first consider the Bragg type of reflection of neutrons on one face (III) of the crystal, i.e. on a family of planes consisting alternately of sodium and chlorine (Fig. 10). First-order reflection occurs when the radiation reflected on the first sodium plane differs in phase by 360° from the radiation reflected on the second sodium plane, by 720° from the radiation reflected on the next sodium plane, etc.

Of course, a phase difference of 360° means that the two beams are in phase. However, there is also a reflection above the intermediate plane of chlorine with a phase difference different from the previous ones for geometrical reasons. In fact, the optical path difference is such that it corresponds to a phase difference of 180°, i.e., a phase inversion: there is therefore phase opposition between the rays coming from sodium and those coming from chlorine. The geometric phase change will therefore make sodium "work" against chlorine; thus if sodium and chlorine scatter neutrons with the same phase, the first order will be weak; on the other hand, if sodium and chlorine scatter with opposite phase, the first order will be strong.

What is the situation like in the second order? In the second order the phase differences from the different optical paths are all doubled i.e. they are 0°, 360°, 720°. Thus in this case there are no phase differences from geometrical reasons. In fact, 360° means zero phase difference, so the opposite case occurs, i.e.: if sodium and chlorine scatter neutrons in the same way, their actions add up; if sodium and chlorine scatter neutrons in the opposite way, their actions subtract. If the intensities of successive orders are observed, in the case that sodium and chlorine diffuse neutrons with equal phase, one will find the first order weak, the second order strong, the third order weak, the fourth order strong, and so on: generally odd orders weak, even orders strong (Fig. 11a).

**Fig. 10**

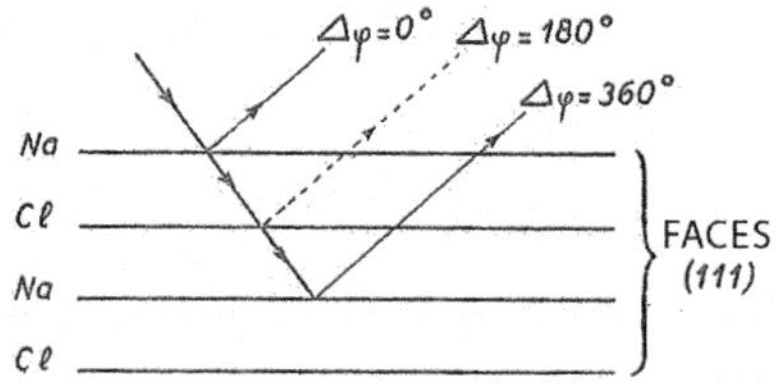


**Fig. 11**

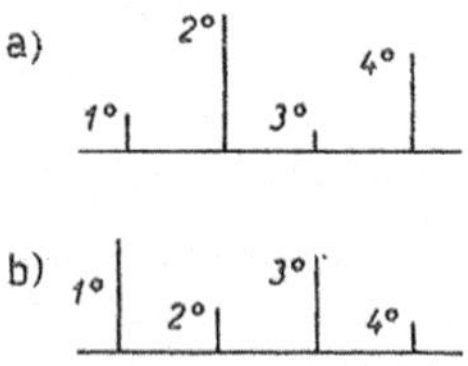

In the case of differentiated phases, on the other hand, the opposite situation will be found, namely, first order strong, second order weak, third order strong, and so on (Fig. 11b). In reality, the situation is a little more complicated, and those familiar with similar experiments performed with X-rays know that, going from order to order, there is a general decrease in intensity due to a number of causes that I will not go into here, so that a trend of the kind described is superimposed on a general decrease in intensity of the low orders toward the high orders.

If experiments are made over face (100), which, as I said before, consist of planes all equal, each containing 50% atoms of one species and 50% atoms of another species, only the general decrease in intensity is found. I want to report some numerical values obtained from measurements made with a crystal of lead sulfide, which has the same crystal structure as sodium chloride. I will first indicate how the experiment was conducted. One mechanically couples the arm holding the detector at $BF_3$, with the platform carrying the crystal so that a rotation of a certain angle of the crystal corresponds to a rotation of twice the angle of the detector. Thus it is certain that the law of reflection (angle of incidence equal to angle of reflection) is always observed for any angle.

Starting from small angles and increasing the rotation toward large angles, one first obtains zero intensity. Then the intensity rises rapidly tending to a maximum which is reached when Bragg's relation for first-order reflection is satisfied. Still continuing the rotation toward increasing values of angles, the intensity decreases until a second maximum is reached at second-order reflection, and so on in succession for successive orders. Here are the values of the intensities (measured in numbers of pulses per minute) observed in the reflection on one face (100) of a crystal of lead sulfide. On that face one should expect, as I have said, only the general decrease in intensity without complication of any other nature.

These values show a decrease quite in agreement with the law of general intensity decrease. In contrast, the following values are found for the reflection on one face (111).

| | | | | |
|---|---|---|---|---|
| Face (100): | 1° | order | 19600 | pulses per minute |
| | 2° | " | 11400 | " " |
| | 3° | " | 2250 | " " |

| | | | | |
|---|---|---|---|---|
| Face (111): | 1° | order | 7300 | pulses per minute |
| | 2° | " | 10700 | " " |
| | 3° | " | 808 | " " |
| | 4° | " | 750 | " " |

It is quite clear from these values that, superimposed on a general decrease, we have the first order weak, the second order very intense (because the increase in intensity of the second order is so great as to counterbalance the general decrease), the third order very weak, and so on. The table clearly shows how odd orders are weak, even orders strong. It is concluded that lead and sulfur scatter neutrons with the same phase.

If one does the same experiment with a crystal of lithium fluoride also having the same structure as sodium chloride (I refer to the structure of NaCl simply because it is very well known, but, of course, one can, with appropriate calculations, extend these considerations to crystals with more complicated structure as well) the reflection on the face (111) gives the following intensities:

| | | | | | |
|---|---|---|---|---|---|
| 1° | order | 10080 | pulses | per | minute |
| 2° | ″ | $\sim 0$ | ″ | | ″ |
| 3° | ″ | 300 | ″ | | ″ |

This is a very significant case in which the situation is completely reversed in that against very intense odd orders one has even orders so weak that they disappear altogether. It is concluded that lithium and fluorine scatter neutrons with opposite phases.

From the systematic study of a large number of crystals it has been found that neutron scattering almost always occurs with the same phase and, to be precise, almost always with positive phase, that is to say the quantity $a$ is positive; which in itself may give rise to considerable surprise when one keeps in mind the previously illustrated geometrical meaning of the quantity a. In fact, in Figs. 2, 3 the straight line of the function $r\psi$ starts quite far from the origin and it is therefore quite improbable that it is so little inclined as to intersect the x-axis on the negative side.

Of course, this is not impossible: the occurrence of either circumstance depends on the particular shape of the $r\psi$ function but it is easy to understand that positive intersections are by far the most likely. In fact such a statistic is found. Out of more than thirty elements studied, only three give opposite phase, namely: lithium, manganese and hydrogen. The latter, however, represents a very complicated case that requires special attention. As is well known, X-rays are used in the investigation of crystals to observe their structure and to determine the position of the various atoms in the elemental cell of the crystal. With X-rays, however, it is not possible to fix the position of the hydrogen atom because it does not produce X-ray refraction, especially if ionized. Crystallographic study with X-rays of hydrogen is therefore impossible. A similar difficulty is naturally not encountered with neutrons because they are scattered not by electrons but by the nucleus, and hydrogen can lose its electrons, but ... if it loses the nucleus, it will also lose itself.

Therefore, from the very beginning of these studies, there was an early conviction that a remarkable field of application could be hydrogen crystallography. Unfortunately, hydrogen crystallography by this method, while not impossible, is nevertheless far from easy. In fact, the early stages were encouraging, but it did not go very far, at least not as far as one might have hoped then. The reason is as follows: the hydrogen nucleus, the proton, like many other nuclei and indeed like the neutron itself, has spin. Consequently (and here I am referring to hydrogen, but what I am saying applies of course to neutron diffusion by any nucleus endowed with spin) it can happen that the neutron after diffusion has its spin oriented in the original direction or in the opposite direction. The change in the angular momentum of the neutron is accompanied, and compensated for, by a change in the angular momentum of the nucleus. If there is no spin reversal, there is coherent scattering; if there is spin reversal, there is incoherent scattering, that is, scattering incapable of giving rise to interference phenomena.

To realize this fact in the simplest way, let us suppose that on a crystal, whose nuclei have their spin oriented in certain ways, a neutron impinges, which is scattered in a certain direction. For there to be interference phenomena it is necessary, so to speak, that all these nuclei cooperate in the phenomenon. Now if the direction of the spin of the scattered neutron is "overturned", there will be correspondingly a reversal of the spin of one of the scattering nuclei. Then this very fact indicates which one is "guilty" of the act of scattering: it was that nucleus that overturned its spin. The phenomenon affects it alone and there is no possibility of interference. If, on the other hand, there is no reversal of the spin of the diffused neutron, it cannot be known which of the nuclei is responsible for the act of diffusion. The responsibility is divided among all of them and interference takes place. Hydrogen has a large cross section, which would in itself be favorable because it would make hydrogen convenient to observe, but unfortunately this large cross section corresponds (for 95%) to phenomena in which the spins flip, i.e., to incoherent scattering phenomena. Only about 5% of the cross section refers to coherent scattering phenomena and can be used to observe the interference phenomenon. The situation for hydrogen is unfavorable in two ways: on the one hand because the intensity of hydrogen contribution is small and thus in itself is difficult to observe, and on the other hand because coherent scattering (interference phenomena) is small against a large background due to incoherent scattering, which makes it even more difficult to observe. This is the reason why studies of identification of the position of hydrogen in crystals by neutron diffraction have not yet made much progress although, as I said, there have been encouraging beginnings; perhaps, little by little, the technique will be refined until they can be used conveniently.

There would be various other semioptical properties of neutrons that I could mention, but since there is not much time I prefer to talk about one in particular and precisely about the phenomenon of neutron reflection, reflection understood not in the Bragg way but, as it were, in the normal sense. It is also observed for X-rays but only under extremely grazing incidence, that is, when the radiation arrives almost parallel to the mirror. The same occurs in the case of neutrons, for the reason I shall now state. Before indicating how the experiment is done I want to point out that neutron reflection is observed only when the incident beam forms an angle with the mirror of the order of a few arcminutes. So to obtain reflection on a mirror of a beam with such a grazing incidence it is necessary to prepare a very strong collimation. In the first experiments the first slit was, let us say, 1 mm wide and the second one, of the same width, was placed 5 m away so that the beam presented an extremely severe delimitation. This condition of course obliges one to use a source of great intensity otherwise one would have to wait a long time before finding the neutron passing through the two slits.

Beyond the second slit (Fig. 12) is the mirror, consisting of a polished sheet of various materials, about 10 cm by 20 cm in size, tilted to the direction of the incident beam by a carefully controlled angle of a few minutes. To observe the reflected beam at a distance of 3 m a boron trifluoride counter was placed, in front of which another slit was made in the cadmium screen, also about 1 mm wide. Finally, the counter was rotated until the reflected beam was captured.

**Fig.12**

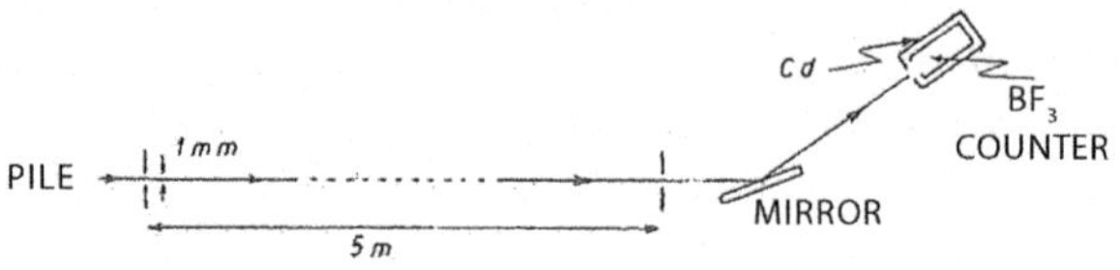


**Fig.13**

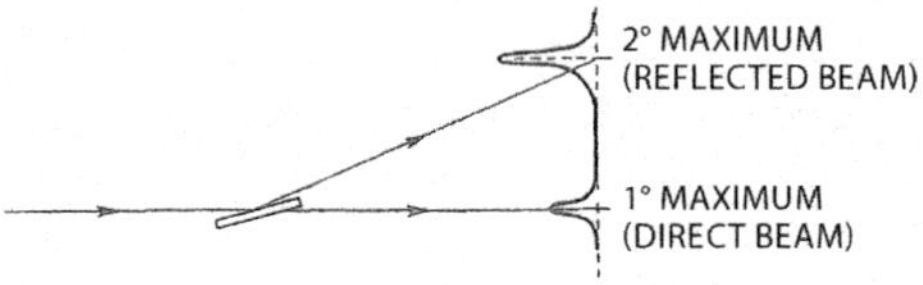


Actually, in such an experiment, due to the intensity of the reflected beam, two maxima are found in two different directions (Fig. 13); the first corresponds to those few neutrons passing directly through the mirror, the second, quite often more intense, corresponds to the reflected neutrons. If one then rotates the crystal, one finds that these two maxima shift in agreement with the laws of reflection.

It is also desirable for the reflection phenomenon to be studied with monochromatic neutrons. This can be achieved by having sufficient intensities and introducing a monochromator before the collimating slits. With monochromatic neutrons the intensity of reflection can be observed as the angle θ by which the crystal rotates varies. Plotting in a diagram the angle θ and in ordinates the intensity of the reflected beam, it is found that the intensity remains almost constant until a certain value of the angle from which it falls abruptly toward zero (Fig. 14).

**Fig.14**

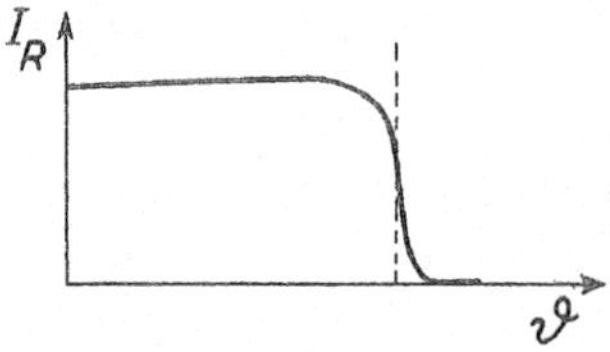


This fact indicates that it is essentially a phenomenon not of ordinary reflection, but of total reflection; in fact it is characteristic of total reflection that as long as one remains below the limiting angle the reflection is strong, as soon as one reaches the value of the limiting angle or goes beyond it by a little the intensity falls to zero very rapidly. The total reflection of neutrons is made possible by the fact that the index of refraction for neutrons, is, in almost all substances, and precisely in normal substances giving 180° phase change, a little less than one, differing by an amount of the order of a few millionths. As a result, the limiting angles are extremely small which leads to extremely difficult geometrical conditions of observation.

What is this refractive index of neutrons due to? Why do neutrons refract? Essentially for the same reason they scatter; in fact, scattering and refraction of light, X-rays, neutrons or any other radiation, are two phenomena that are always very intimately related. Consider a wave and then interpose in its path a small plate of any material—the index of refraction appears as a phase loss, or rather as a phase change. If the index of refraction is, for example, greater than one, in the interior the wavelength is somewhat shorter, so that at the exit the wave will be found with the phase shifted. The connection between the phenomenon of diffusion and that of refraction can also be put into quantitative form by means of the following relationship:

$$n = 1 - \frac{\mathrm{N}a}{2\pi}\lambda^2 \tag{5}$$

where N is the number of atoms per $\mathrm{cm}^3$, $a$ is the usual constant already introduced which is characteristic of diffusion and $\lambda$ is the de Broglie wavelength. The term $(\mathrm{N}a/2\pi)\ \lambda^2$ is of the order of a few millionths. From the above formula it can be seen that if $a$ is positive the index of refraction is less than one; if $a$ is negative it is greater than one.

But since for the reasons already seen, and as is moreover evident from experiments, in almost all substances, (to be precise in about 90% of substances) a is positive, a refractive index less than one is to be expected in most cases. One can use the previous expression of the refractive index to find the limiting value $\theta_0$ (complementary of the angle of incidence). As a first approximation one finds for the limiting angle $\theta_0$ the expression:

$$\theta_0 = \sqrt{2(1-n)} = \lambda\sqrt{\frac{\mathrm{N}a}{\pi}} \tag{6}$$

or even, remembering $a=\sqrt{\sigma/4\pi}$:

$$\theta_0 = \lambda\sqrt{\frac{\mathrm{N}}{\pi}\sqrt{\frac{\sigma}{4\pi}}} \tag{7}$$

This formula is amenable to experimental verification, because the wavelength is known, since one can work with monochromatic neutrons, and the values of N and $\sigma$ are also known.

I will conclude by reproducing a table in which the values of $\theta_0$ calculated as above and the values of $\theta_0$ measured experimentally on mirrors of various kinds appear. For example, a beryllium mirror was used, which is a nice metal, easy to clean, for which the values of $\theta_0$ appear to be quite close. Then the experiment was carried out with a graphite mirror, which does not take cleaning well but requires a great deal of work without ever turning out very nicely. So that the experiment does not turn out to be very accurate, resulting in quite a large difference between the theoretical and experimental values of $\theta_0$. Later other mirrors such as iron, nickel, zinc and copper were tried, the data of which are collected in the table below:

| Mirror | | $\theta_o$ measured | $\theta_o$ calculated |
|---|---|---|---|
| Be | .......... | 12,0′ | 11,1′ |
| C | .......... | (10,5)′ | 8,4′ |
| Fe | .......... | 10,7′ | 10,0′ |
| Ni | .......... | 11,5′ | 11,8′ |
| Zn | .......... | 7,1′ | 6,9′ |
| Cu | .......... | 9,5′ | 9,5′ |

From here we can see that in general the agreement is quite good. This fact indicates that there is a very good correlation between theory and experiment.

There would be in this order of considerations many other phenomena to illustrate but unfortunately I do not have the time. Instead I would like to make a few remarks on the results of this work, the purpose of which is perhaps to a great extent more didactic than scientific. Quantum mechanics itself is in fact now well established. No one in fact today could doubt de Broglie's relation that gives the wavelength associated with corpuscles. However, it is interesting to test it in a big way because, after all, it is a theory that is basic to our perception of physics. Educational experiments of this kind require a pile (and there are few piles!) so it would be difficult to bring them before an audience such as this one. When you can actually do it however seeing experiments such as these that can be assembled with great ease is really very instructive.

As far as the scientific value of this research is concerned, it can be noted that probably the field in which they will be able to have some consequences is somewhat in competition with X-rays, because of the problem of crystal structure. From this point of view, in general, the advantage is all on the side of X-rays, for two reasons.

1st X-rays can be obtained with much stronger intensity than neutrons;

2nd an X-ray apparatus is a very common apparatus, while a pile is a very expensive apparatus of which there are few specimens in the world.

There are some cases, admittedly not many, in which the advantage is on the side of the neutrons. One of these is the case I mentioned earlier and, that is the case of hydrogen, on which, while known X-rays can do absolutely nothing, neutrons can instead work, albeit, at least for now, with difficulty. Certainly, however, little by little, the difficulties will be overcome. It is a matter of increasing the intensity and the power of resolution. These are all problems that can be solved. There are also other cases in which X-rays are rather at a disadvantage. Suppose, for example, we have an alloy of Cu and Zn; these metals are contiguous in the periodic system, so they have approximately the same number of electrons. Therefore, X-rays, which are scattered by electrons do not "distinguish," so to speak, copper from zinc, which neutrons instead can do. They "feel" the nuclear properties of Cu and Zn nuclei between which there is no similarity. In a problem, for example, of order and disorder in a Cu and Zn alloy, X-rays do not notice whether the two elements have been exchanged, while neutrons immediately notice the difference. There are thus a number of applications, not spectacular I would say but also of some interest, to which this method lends itself and will increasingly lend itself in future.

## Biographical Notices on Carlo Salvetti

### AT and PT

Carlo Salvetti was born in 1918 in Milan, where, in 1940, he graduated in Physics. There he was a student of Giovanni Gentile jr (1906–1942) who, in that period, taught Probability theory and Metal theory.

In 1941–43 he served in the Army. After 8 September 1943 he refused to join the Republic of Salò from where Mussolini hoped to reconstitute the Fascist regime and went into hiding.

At the end of the war in 1945 he began his research in the field of cosmic rays. Together with Giorgio Salvini (1920–2015), both assistants to Giuseppe Bolla (1901–1980), he was involved in the task of studying the possible developments of the peaceful use of nuclear energy. Mario Silvestri (1919–1994) and Salvetti drew up a research program for the construction of a zero-power atomic pile similar to Enrico Fermi's CP-1 in Chicago (1942) and the construction of a small power reactor.

The idea of setting up of a new center arose in this context.The new institution was founded in 1946 and was called Centro Italiano di Studi ed Esperienze CISE (Italian Centre for Studies and Experiments), a private research institute on nuclear physics and technology. In the early '50 s he was a visiting Professor in India, assisting and guiding the Indian Atomic Authority in the development of a national chain of heavy water reactors.

In 1952 the Consiglio Nazionale delle Ricerche (CNR) (National Research Council) established the Comitato Nazionale per la Ricerca Nucleare (CNRN) (National Committee for Nuclear Research) which in 1953 provided funding and personnel for the Frascati Electrosynchrotron, completed in 1959.

In 1956 the Ispra Research Centre was established, the first CNRN research infrastructure, and a 5MW CP5 experimental nuclear reactor was ordered. Salvetti was the director.

In 1960 the Italian Parliament transformed the CNRN into the Comitato Nazionale per l'Energia Nucleare (CNEN) (National Committee for Nuclear Energy). Felice Ippolito (1915–1997) continued to serve as Secretary General of CNEN. In the same year the Italian Government sold (on rather disadvantageous terms) the Ispra Research Centre to EURATOM and the CP5 reactor only a few months after its completion. The negotiation was conducted in a non transparent manner and Salvetti resigned.

In 1959 he became Director of Research at the Laboratories of the International Atomic Energy Agency (IAEA) in Vienna.

In 1963, following the defenestration of the director Ippolito for minor administrative irregularities, followed by a trial that imposed a heavy sentence, Salvetti was appointed vice-president of CNEN, replacing Ippolito.

He was appointed to a chair in Bari (1954), and later taught in Milan (1959) and in Rome in 1981.

He died in 2005.

For further informations, see: Bonolis (2005); Gariboldi et al. (2022); Lavrencic and Valli (2005); Salvetti (1970).

## References


1. Bonolis, Luisa (2005). "Giovanni Gentile Jr a Milano" Atti del XXV Congresso Nazionale di Storia della Fisica e dell'Astronomia, Milano, 10-12 novembre 2005 (Milano: SISFA, 2005)
2. Gariboldi, Leonardo; Bonolis, Luisa; Testa, Antonella (2022). The Milan Institute of Physics. A Research Institute from Fascism to the Reconstruction (Cham: Springer International Publishing, 2022).
3. Lavrencic, Devana; Valli, Giulio (2005). "In ricordo del prof. Carlo Salvetti" Energia, Ambiente e Innovazione 2005 51: 23-24.
4. Salvetti, Carlo (1970). "Nuclear energy in Italy" https://inis.iaea.org/collection/NCLCollectionStore/_Public/01/000/1000475.pdf, accessed in September 28, 2023

# Nineth Lecture The Dirac Monopole

**Enrico Fermi**

## Edited by P. Caldirola

About twenty years ago, the English theoretical physicist Dirac[1] published some speculations on the possibility that there exists a particle endowed with a magnetic charge instead of an electric charge: in other words, an isolated magnetic pole, a monopole. This is a particle that no physicist has so far "seen" but whose existence is, at least theoretically, possible. Now my interest in this particle was revived, about a couple of years ago, by the arrival of the first (incomplete, fragmentary and largely erroneous) news about the discovery of heavy particles in cosmic radiation. To be precise, it was learned that, on photographic plates sent to great heights, particles had been observed whose ionizing power appeared to be much more intense than that of light particles or the usual heavy particles such as protons and $\alpha$-particles; and at first, in some discussions with my colleague Edward Teller, the possibility was examined that these particles instead of heavy nuclei were Dirac monopoles.

When the reports of the experiments performed were later published, it could be quickly recognized that the interpretation originally given was correct, namely, that these were indeed heavy nuclei or at least not light nuclei, which apparently are part of the primary component of cosmic radiation.

But at that time, for a few days at least until the matter was fully cleared up, we became interested in the monopole by trying to understand its behavior.

[1] The magnetic monopole theory was formulated by P. A. M. Dirac in "Proc. Roy. Soc." A, 133, 60 (1931) and was taken up by the same author in a recent Note published in "Phys. Rev." 74, 817 (1948).

(*) Held on October 21, 1949 in the assembly hall of the Montecatini Company in Milan.

E. Fermi (✉)
University of Chicago, Chicago, IL, USA



A. Treves and P. Tucci (eds.), *Enrico Fermi, Atomic Physics Lectures*, History of Physics,
https://doi.org/10.1007/978-3-031-68690-0_12

In this lecture I would like to entertain you a little about the nature and properties of this hypothetical particle. This, perhaps, is not entirely idle because, if the particle in question were to actually exist, it might someday occur to someone to observe it: and this might happen in a country, such as Italy, where cosmic ray studies are flourishing so much. It is therefore worthwhile to try to realize, from a general point of view, the characteristics and effects that such a particle should present.

I will begin with a preliminary consideration, which is as follows: it is well known that the electric field and the magnetic field are two entities with very similar properties. In other words, the equations obeyed by the electric field in vacuum are almost identical to the corresponding equations obeyed by the magnetic field. In the considerations then made about the symmetry properties of these two fields, it is customary to say that the electric field is represented by a polar vector and the magnetic field by an axial vector.

We try to illustrate this in the following way: in a reflection of space with respect to a point (an operation that from the physical point of view cannot of course be performed) a polar vector is a vector that—such as for example the displacement vector from a point A to a point B—changes direction, while an axial vector is a vector that preserves, in such an operation, its sign unchanged. Actually, since the operation of spatial reflection cannot be carried out, the polar and respectively axial nature of the electric field and the magnetic field is established by considerations of various kinds. If, for example, we suppose that we have (Fig. 1) in O a proton and in A an electron, the electric field is directed from the proton to the electron, and, if we reflect space, the electron comes to be in A' and the electric field therefore changes direction.

**Fig.1**

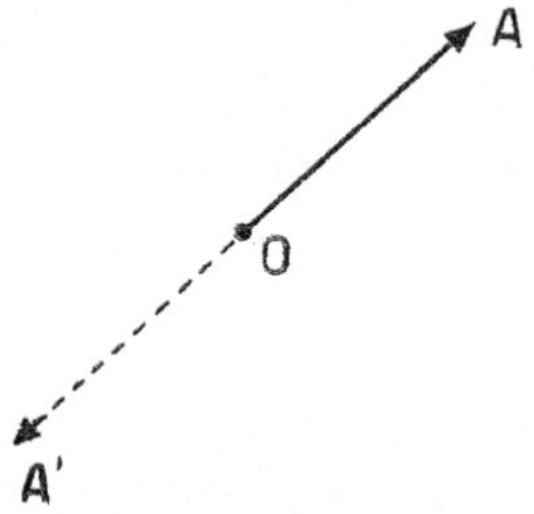


If we admit the existence of an isolated magnetic monopole, we could conclude, with an argument of the kind now given, that the magnetic field can also be represented by a polar vector. The difference in symmetry between the electric and magnetic fields would thus be disturbed, unless of course, alongside our pole, we always admit the existence of another pole of equal mass and opposite sign.

Dirac for the first time posed the problem of the possibility of existence, from the point of view of quantum mechanics, of an isolated magnetic pole arriving at a result that is not entirely obvious a priori: *the magnetic mass of the supposed monopole cannot have an arbitrary value but must be an integer multiple of a certain quantum of magnetism*!

I would like to present now Dirac's argument in a somewhat simplified way. Suppose we have at a certain point P in space a monopole of magnetic mass $m$. It would give rise to a magnetic field analogous to the electric field generated by an electric charge. In other words, at a distance r from the monopole we would have a magnetic field $\overrightarrow{\mathrm{H}}$ given by:

$$\overrightarrow{\mathrm{H}} = \left(\frac{\overrightarrow{m}}{r^2}\right).$$

Such a field obviously would not have zero divergence, given the existence of a source in P. Describing, as usual, the magnetic field H through the vector potential $\overrightarrow{\mathrm{A}}$, we will have[2]:

$$\overrightarrow{\mathrm{H}} = \mathrm{rot}\,\overrightarrow{\mathrm{A}},$$

with $\overrightarrow{\mathrm{A}}$ a vector with zero divergence:

$$\mathrm{div}\,\overrightarrow{\mathrm{A}} = 0.$$

The potential vector $\overrightarrow{\mathrm{A}}$ is completely determined by these two equations, which mathematics teaches how to solve. To avoid explicit calculation, however, we will achieve the same aim by making use of an electrical analogy.

Suppose we have a system of density currents $\overrightarrow{\mathrm{j}'}$, which gives rise to a magnetic field $\overrightarrow{\mathrm{H}'}$ such that

$$\begin{cases} \mathrm{div}\,\overrightarrow{\mathrm{H}'} = 0 \\ \mathrm{rot}\,\overrightarrow{\mathrm{H}'} = 4\pi j'. \end{cases}$$

Comparing these equations with the previous ones to which the vector satisfies $\overrightarrow{\mathrm{A}}$ we see that the problem of determining the vector potential given the magnetic field $\overrightarrow{\mathrm{H}}$ is mathematically equivalent to the problem of determining the magnetic field $\overrightarrow{\mathrm{H}'}$ produced by a system of density currents $\overrightarrow{\mathrm{j}'}$. The correspondence between these two problems can be summarized in the following table

| Original problem | Equivalent problem |
|---|---|
| given: $\overrightarrow{\mathrm{H}}$ | given: $4\pi\,\overrightarrow{\mathrm{j}'}$ |
| sought: $\overrightarrow{\mathrm{A}}$ | sought: $\overrightarrow{\mathrm{H}'}$ |

Since the way to calculate the magnetic field of a current by means of the various laws of Laplace, Ampère, etc. is well known and familiar so the original problem is also substantially solved.

[2] The relation $\overrightarrow{\mathrm{H}} = \mathrm{rot}\,\overrightarrow{\mathrm{A}}$ must of course be assumed to be valid at all points in space, except at the point P where the magnetic mass m "source" of the field is located. At that point it is obviously $\mathrm{div}\,\overrightarrow{\mathrm{H}} \neq 0$.

In our analogy we will have to have a system of currents exiting from a point so that at a certain distance r from the source in P we have

$$4\pi \vec{j}' = \left(\frac{\vec{m}}{r^2}\right)$$

Therefore, the point P, where the monopole source of a constant magnetic field flux is located, becomes, in the analogy, the source of a constant current density flux $\vec{j}'$. It is therefore necessary to make electric charge flow continuously into the point P. This can be thought of as being accomplished by means of a conducting wire of arbitrary shape which conveys into P a concentrated current of intensity I equal to that which exits from this point radiating symmetrically in all directions,[3] and thus in the equivalent problem div $\vec{j}' \neq 0$.

It will therefore have to be:

$$\mathrm{I} = \Phi(\vec{j}\,) = \vec{j}' \cdot 4\pi r^2 = \frac{m}{4\pi r^2} \cdot 4\pi r^2 = m$$

where $\Phi(\vec{j}')$ denotes the outgoing vector flux from a surface enclosing the point P. Since, however, the choice of the conductor carrying the current I into P is arbitrary, because this can be done in an infinite number of ways, it also correspondingly results in an infinite number of solutions of the original problem.

**Fig.2**

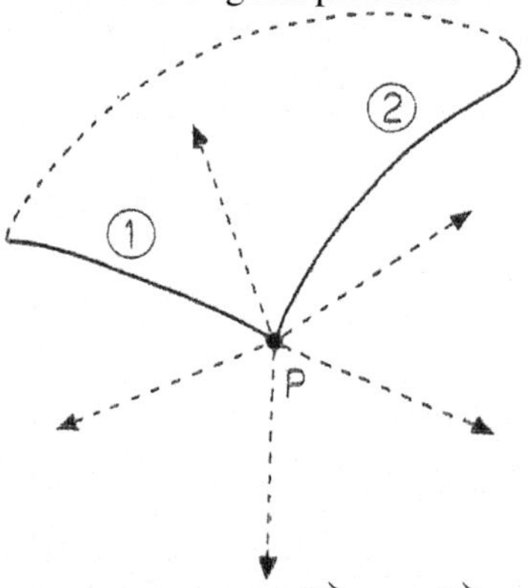


Now (Fig. 2), if we call two of these solutions $\vec{A}_1$ and $\vec{A}_2$, corresponding respectively to the introduction into the equivalent problem of either wire (1) or wire (2), the difference $\vec{A}_1 - \vec{A}_2$ will be numerically equal to the magnetic field $\vec{H}'$ that would be produced by the current distribution that results from the difference between the total currents passing through P when we assume the existence of wire (1) and those passing through it when we assume the existence of wire (2). These systems of currents evidently differ only in the current flowing into P along (1) in the first case

[3] The introduction of these ideal current-inducing wires at point P is required by the fact that there, as we have previously noted, we have div $\vec{H} \neq 0$.

and along (2) in the second. Therefore, with $\vec{H}'(1)$ the field generated by the current flowing into P along (1) and $\vec{H}'(2)$ that generated by the current flowing into it along (2), we shall have:

$$\vec{A}_1 - \vec{A}_2 = \vec{H}'(1) - \vec{H}'(2);$$

in other words, the difference between the potential vectors $\vec{A}_1 - \vec{A}_2$ turns out to be equal to the magnetic field generated by a current that, coming from infinity along (1), reaches P and then leaves it by returning to infinity along (2). Now it is known from electrology that such a magnetic field can be represented by means of a potential that is not a univocal, but a multi-valued function, namely, a function $f$ whose value at a point in space changes by $4\pi$ times the value of the current intensity each time it turns around the wire within which the current passes. We can thus say that:

$$\vec{A}_1 - \vec{A}_2 = \text{grad } f,$$

where therefore $f$ is a potential with infinite values, which each time it passes around the line (1–2) changes by $4\pi m$ (recall in fact that $I = m$). It is precisely from this property of f that Dirac's relation, mentioned at the beginning of this lecture, results, which gives as possible only certain very particular values of the magnetic pole $m$. In fact, it is well known that in quantum mechanics, the wave function $\psi$, which characterizes a particle of charge e, changes phase when the definition of the vector potential of the field acting on it by a gradient is altered; thus if

$$\vec{A} \to \vec{A} + \text{grad } f,$$

$\psi$ is transformed as follows

$$\psi \to \psi \exp\left(\frac{ie}{\hbar c} f\right);$$

That is to say, by turning around the line (1–2), $\psi$ is multiplied by a factor

$$\exp\left(\frac{ie}{\hbar c} 4\,\pi m\right).$$

But this expression cannot have a value other than 1, otherwise the foundations of wave mechanics itself would be upset (for it is well known that it is required that $\psi$ must always be a single valued function). So it will have to be:

$$\exp\left(\frac{ie}{\hbar c} 4\pi m\right).$$

The most general complex number that is equal to 1 is $\exp(2\pi i n)$, where $n$ is an arbitrary integer; therefore

$$\frac{ie}{\hbar c} 4\pi m = 2\pi i n$$

That is,

$$m = n\frac{\hbar c}{2e} = n\frac{\hbar c}{2e^2} \cdot e,$$

where $\hbar c/e^2$ is the well-known number "137" (inverse of the fine structure constant). Thus we have

$$m = n\,\frac{137}{2}\,e = n\,68,5\,\cdot e\,,$$

that is to say, the monopole can only have a magnetic mass that is an integer multiple of the elementary electric charge multiplied by 137/2 = 68.5.

This conclusion can also be reached in another way, which I would now like to briefly expound without going into detail. One of the essential facts of quantum mechanics is that areal momentum is quantized, and to be precise it can take only multiple values of $\hbar/2$. Consider (Fig. 3) a monopole $m$ and an electric charge $e$, as in the figure; at point P it will be:

$$\begin{cases} \vec{\mathrm{H}} = \frac{\vec{m}}{r_2^2} \\ \vec{\mathrm{E}} = \frac{\vec{e}}{r_1^2} \end{cases}$$

Now, the vector product $\vec{E}$ times $\vec{H}$ (which becomes a vector perpendicular to the plane of the figure) represents, divided by $4\pi c$, the electromagnetic momentum density.

**Fig.3**

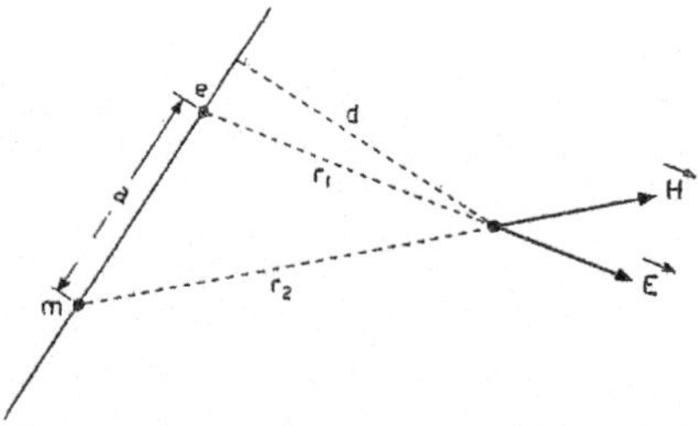


Reasoning roughly about orders of magnitude, if $a$ is the distance between the monopole and the charge it will be:

$$\mathrm{E} \simeq \frac{e}{a^2}, \qquad \mathrm{H} \simeq \frac{m}{a^2}, \qquad \left|\frac{\vec{\mathrm{E}} \wedge \vec{\mathrm{H}}}{4\pi c}\right| \simeq \frac{em}{a^4 c}$$

The "arm" $d$ will also be of the order of a and the "effective" volume of integration will be of the order of $a^3$, so the integral that gives the total momentum has the order of magnitude

$$\frac{em}{a^4c}aa^3 = \frac{em}{c};$$

that is to say, a result is obtained that does not depend on the distance between pole and charge.

The quantization condition then gives:

$$\frac{em}{a^4c} \cdot a \cdot a^3 = \frac{em}{c};$$

which is the relationship derived above.

Now Dirac made a rather curious observation and it is this. A well-known (and for the moment certainly unexplained) fact lies in the circumstance that there is a quantum of electric charge, i.e. the charge of all particles that we know is always an integer multiple of an elementary value which is the charge of the electron. If we admit that at least one monopole exists in the universe we must have:

$$e = \frac{\hbar c}{2m}n,$$

that is, the electric charge will turn out to be quantized; what is more, the magnetic mass of the monopole will have to be assigned a value such that the electric charge consequently assumes the values that are found experimentally. In other words, the value of the elementary electric charge, which we know from endless experiments, "forces" the monopole, if it exists, to have a quantized magnetic mass and vice versa.

In electrodynamics, all radiative properties of the electron are essentially determined by the fine structure constant

$$\frac{e^2}{\hbar c} = \frac{1}{137}.$$

Now, as we said in the first part of this course,[4] the circumstance that this fine structure constant is very small is the one that makes it possible to separate in a first approximation the radiative field from the electric charge. In other words, the fact that the above-mentioned parameter is very small with respect to unity makes it possible to adopt a procedure of calculation by successive approximations in which, in a first approximation, the electron and the electromagnetic field are considered separately as not interacting with each other. Radiative phenomena, in which the two fields (that of the electron and that of the external field) come to "mix," are presented in a higher approximation. For the corresponding phenomena related to the existence of the monopole, the analogous constant of that of fine structure would be

[4] See the 6th lecture.

$$\frac{m^2}{\hbar c} \sim 34,$$

i.e., a rather large number with respect to unity, which means that separating the electromagnetic field from the monopole itself is by no means a correct approximation. It follows that if the monopole exists it will have to behave in a quite different way than should be expected for a magnetic entity purely analogous to an electric charge.

However, without caring about these difficulties, let us go on to discuss what a monopole would look like supposing, for example, that someday someone happened to observe it in cosmic radiation or as a product of reactions obtained through the use of some giant cyclotron. First of all, it would ionize (although its charge is a magnetic charge) in a way not too different from an electric charge, but with some peculiarities that I now wish to mention.

Suppose the monopole travels according to the line drawn in Fig. 4: Let $b$ be the minimum distance at which it passes by an electron $e$.

**Fig.4**

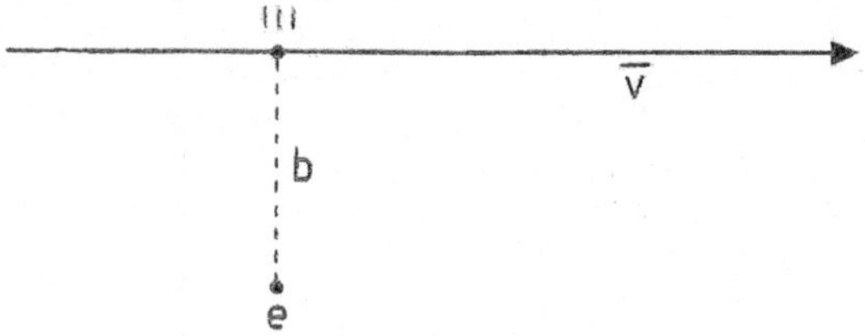


The electron will be in a magnetic field H $\backsimeq m/b^2$; but, as is well known, a magnetic field generated by a particle in motion presents itself to a "stationary" observer as an electric field of intensity

$$\mathrm{E} \backsimeq \frac{m}{b^2}\frac{v}{c}\cdot$$

Note that if a charge $e$ passed with velocity $v$, the electric field generated by it would simply be

$$\mathrm{E} \backsimeq \frac{e}{b^2},$$

and since the ionization effect is essentially determined by the square of the E field, we have:

$$\frac{\text{monopole's effect}}{\text{electron's effect}} \backsimeq \left(\frac{m}{e}\right)^2 \left(\frac{v}{c}\right)^2,$$

and for $v \backsimeq c$ the ratio holds $\sim (m/e)^2 \simeq 68, 5^2 = 4692$, that is to say, the monopole ionizes as much as a nucleus of atomic number Z $\simeq$ 68. And this is the reason why, when heavy nuclei were discovered in cosmic radiation, it was thought that they might be monopoles.

**Fig.5**

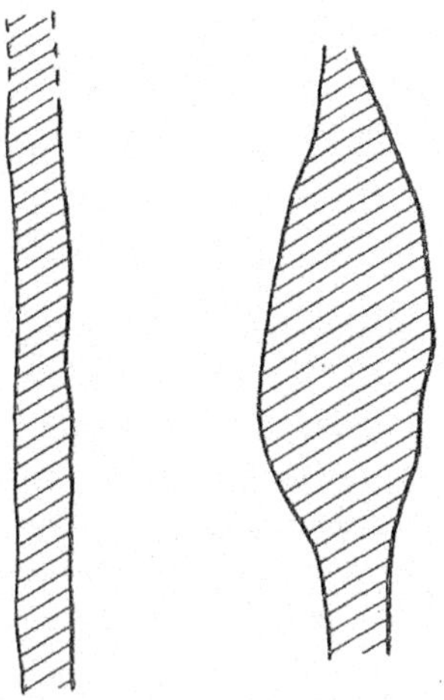

We observe again that at large velocities, the ionization of an electric charge is constant, while at low velocities it is roughly proportional to $1/v^2$; in the case of the monopole, on the other hand, the factor $v^2$ in the numerator eliminates the velocity dependence, resulting in ionization that is practically constant as velocity varies.

In a photographic plate, the trace of a monopole and the trace of a heavy nucleus of $Z = 68$ should have "roughly" the appearance drawn in Fig. 5. Toward the end, the nucleus trace becomes thin, because the nucleus begins to capture electrons, thus causing a decrease in charge and thus ionization. The monopole track, on the other hand, would remain substantially uniform, apart from the final extreme section, the appearance of which is very difficult to predict. Any monopole contained in cosmic radiation would then be affected by the Earth's magnetic field, of course in a different way than nuclei are affected. In fact, the Earth has a magnetic field whose lines of force are arranged roughly as in Fig. 6.

**Fig.6**

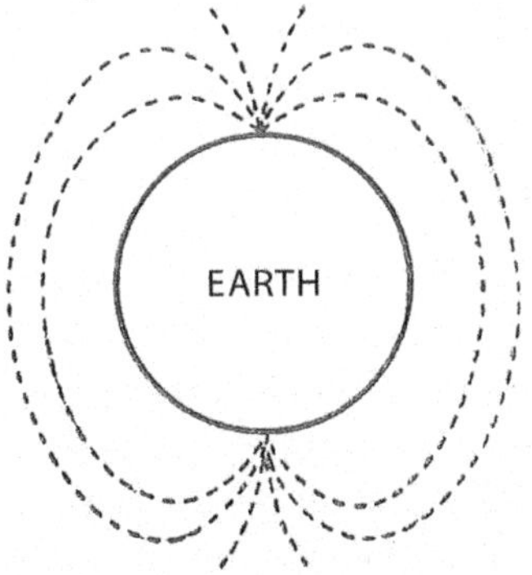


If we liken the Earth's magnetic field to that generated by a dipole of moment $\mu$, the magnetic potential energy of the monopole would be, if $\theta$ is the colatitude:

$$W_\circ = 68e\frac{\mu}{r^2}\cos\theta;$$

in particular at the pole

$$W_{\circ} = \frac{\mu}{r^2}\, 68\, e \backsimeq 5 \cdot 10^{12}\, eV \cdot$$

Let us now admit that a monopole that has entered the Earth's atmosphere has lost its energy through ionization. However, the earth's magnetic field would always tend to make it flow through the air, and, depending on whether it loses energy more or less rapidly than it gains it, it would either be accelerated or move by viscous motion. Well, it is easily found that, for air at its natural density, the critical value of the field is about 10 gauss, and that is to say, a field more intense than 10 gauss is sufficient to make the monopole overcome the resistance of the air and to impart to it an accelerated motion. A weaker field would instead impart to it a uniform motion.

Since the Earth's magnetic field is about 1/2 gauss, it is therefore about 20 times too weak to accelerate the monopole. It would seem, therefore, that the hypothetical monopoles should reach the Earth moving in viscous motion. The positive ones would fall in the southern hemisphere and the negative ones in the northern hemisphere. They would thus tend, if they arrived in large numbers, to neutralize the Earth's magnetic field.

It must therefore be concluded from the absence of such an effect that even if monopoles exist in nature they are very limited in number. On the other hand, however, the most important consequence of their possible existence consists in the fact that the existence of even one of them is enough to impose the quantization of electric charge.

At the end of these lectures, I would like to thank all my listeners and express my gratitude to the Donegani Foundation of the Accademia dei Lincei and the Montecatini Society, whose generous contributions have made this very pleasant stay of mine in Italy possible.

## Biographical Notices on Piero Caldirola

### AT and PT

Piero Caldirola was born in Como in 1914. After high school studies in the scientific field he was admitted to the Collegio Ghislieri and graduated in physics at the university of Pavia in 1937, discussing a thesis on the diffusion of hydrogen flowing through heated palladium. In the same year he joined the Fermi group in Rome.

He was introduced to Fermi by Rita Brunetti (1890–1943), professor of Physics at Pavia University. After Fermi's departure he followed Gian Carlo Wick (1909–1992) to Padua, where he worked on Rabi's theory of the method for the measurement of nuclear magnetic moments, carrying out an investigation of non adiabatic processes in an oscillating magnetic field. Then he returned to Pavia where he collaborated with Luigi Giulotto (1911–1986) in studies on Raman spectroscopy in calcite and carbon bisolphid. In 1941 he published a study on non conservative forces in quantum mechanics which led to the Caldirola-Kanai model of quantum dissipative systems.

He was nominated full professor in 1947. He moved to Milan in 1950 but he did not abandon Pavia. The Pavia and Milan groups, under his direction, gave birth to the first studies in theoretical solid state physics in Italy. In Milan Caldirola carried out research on the magnetic moment of a deuton, on a theory of the meson component in cosmic radiation, on the relativistic equation of motion of an electron in classical electrodynamics, on ergodic methods in statistical mechanics, on the theoretical study of isotope separation. The results of this latter research were patented and used not only by the Centro Informazioni Studi ed Esperienze (CISE) (Information Centre Studies and Experiments) but also by the French Commissariat è l'Energie Atomique at the Nuclear Site of Tricastin in the Pierrelatte nuclear plant.

In 1966–'67 he became president of the committee of CNEN for studies on nuclear fusion at the Laboratories for Ionized Gases in Frascati and gave birth to the group of plasma physics at the University of Milan.

He was a member of the board of directors and consultant of AGIP Nucleare from 1958 to 1963 and of ENI from 1960 to 1963. He was also a member of the scientific and technical committee of Euratom from 1968 to 1973 and chairman of the study group for uranium enrichment of CNEN (National Committee for Nuclear Energy) from 1968 to 1978.

Moreover he was a member of various commissions of NATO from 1965 to 1975 for the supervision of the scientific activities of the laboratories in the Mediterranean area. This multiplicity of theoretical interests, management activities and of curiosity for modern art constituted an alternative vision with respect to the widespread culture in Italian physics, polarized mainly towards research on high-energy particles.

He died in 1984.

For further informations, see: Belloni (1988); Galdabini (1983), and Gariboldi et al. (2022).

## References


Belloni, Lanfranco (1988). "Caldirola, Piero" in Dizionario Biografico degli Italiani (Roma: Istituto della Enciclopedia italiana, 1988), vol. 34.

Galdabini, Silvana; Giuliani, Giuseppe (1983) "Intervista a Piero Caldirola" in Bonizzoni, Ilaria (2002). "Le interviste" in: Giuliani, Giuseppe (ed) (2002). Per una storia della fisica italiana: 1945 - 1965. Fisica della materia, fisica teorica, insegnamento della Fisica (Pavia: La Goliardica Pavese, 2002).

Gariboldi, Leonardo; Bonolis, Luisa; Testa, Antonella (2022). The Milan Institute of Physics. A Research Institute from Fascism to the Reconstruction (Cham: Springer International Publishing, 2022).

The manufacturer's authorised representative in the EU is Springer Nature Customer Service Centre GmbH, Europaplatz 3, 69115 Heidelberg, Germany. If you have any concerns regarding our products, please contact ProductSafety@springernature.com

Printed and bound by CPI Group (UK) Ltd, Croydon, CR0 4YY

30/09/2026

02225604-0012